Spectral Methods in Transition Metal Complexes

Spectral Methods in Transition Metal Complexes

K. Sridharan
School of Chemical and Biotechnology
SASTRA University
Thanjavur
Tamil Nadu
India

ELSEVIER　AMSTERDAM • BOSTON • HEIDELBERG • LONDON • NEW YORK • OXFORD
PARIS • SAN DIEGO • SAN FRANCISCO • SINGAPORE • SYDNEY • TOKYO

Elsevier
Radarweg 29, PO Box 211, 1000 AE Amsterdam, Netherlands
The Boulevard, Langford Lane, Kidlington, Oxford OX5 1GB, UK
50 Hampshire Street, 5th Floor, Cambridge, MA 02139, USA

Notices
Knowledge and best practice in this field are constantly changing. As new research and experience
broaden our understanding, changes in research methods, professional practices, or medical treatment
may become necessary.

Practitioners and researchers must always rely on their own experience and knowledge in evaluating
and using any information, methods, compounds, or experiments described herein. In using such
information or methods they should be mindful of their own safety and the safety of others, including
parties for whom they have a professional responsibility.

To the fullest extent of the law, neither the Publisher nor the authors, contributors, or editors, assume
any liability for any injury and/or damage to persons or property as a matter of products liability,
negligence or otherwise, or from any use or operation of any methods, products, instructions, or ideas
contained in the material herein.

British Library Cataloguing in Publication Data
A catalogue record for this book is available from the British Library

Library of Congress Cataloging-in-Publication Data
A catalog record for this book is available from the Library of Congress

For information on all Elsevier publications
visit our website at http://store.elsevier.com/

ISBN: 978-0-12-809591-1

Working together
to grow libraries in
developing countries

www.elsevier.com • www.bookaid.org

CONTENTS

LIST OF FIGURES

LIST OF TABLES

PREFACE

It is very important to determine the structure of compounds in order to predict and understand their properties so that they can be used as drugs, materials of importance, catalysts, etc. Natural products belong to another important class of compounds in the field of drug discovery. They are isolated from natural sources such as plants, animals, ocean, etc., with great difficulty. Sometimes when it is started with thousands of kilograms of leaves, barks, roots, flowers, etc., we may end up with a few milligrams of the active compound. Hence, we cannot afford to waste this precious compound in order to determine its structure. Therefore, spectral techniques are very valuable in structure determination. These techniques do not destroy the compounds and also the results are obtained quickly. Thus "spectral methods of identification of organic compounds" are widely used.

Just like organic compounds, metal complexes are also widely used as anticancer drugs, catalysts, sensors, labels in MRI techniques to detect various diseases, etc. Spectral techniques such as UV-Vis, IR, NMR, and EPR are widely used in determining their structures. While a large number of books are available for organic compounds, books on spectral methods for inorganic compounds and complexes are scarce. In particular, all these methods are discussed in this single book. This book has taken shape after about three decades of teaching. I hope that this book may be very useful for students and teachers.

I am quite fascinated by Cardinal Newman's words quoted in I.L. Finar's book *Organic Chemistry* (Longman's, 1951), "A man would do nothing, if he waited until he could do it so well that no one would find fault with what he has done." Hence, I welcome suggestions and criticisms on this book.

K. Sridharan
SASTRA University
November, 2015

ACKNOWLEDGMENTS

A person may have potential. But he/she can perform only when a stage is provided. In this regard, I thank SASTRA University for providing me with a stage to perform. I sincerely thank the Vice-Chancellor of SASTRA University, Prof. R. Sethuraman, for his constant support, encouragement, and the excellent academic atmosphere provided. I thank Dr. S. Vaidhyasubramaniam, Dean (Planning and Development) and Dr. S. Swaminathan, Dean (Sponsored Research and Director, CeNTAB) for their constant support. I also thank my students for giving me an opportunity to teach this subject for about 37 years. I thank my parents for their blessings. Last but not least, I thank my wife and sons for their patience and support.

CHAPTER 1

The Electromagnetic Spectrum

1.1 TRANSITION METAL COMPLEXES

Transition metals are the d-block elements and they have incompletely filled d-orbitals. In other words, they have d^1 to d^9 electrons. The d^{10} metals, namely, Zn, Cd, and Hg, have completely filled d-orbitals. However, they are also considered as transition metals because they have similar properties to those of transition metals. When these transition metal ions react with a species, organic or inorganic, capable of donating a pair of electrons, a coordinate bond is formed between the metal ion and the electron donating species. The resulting compound is called a transition metal complex. In other words, the complex may be considered as a product of a Lewis acid-base reaction. The metal ions are the Lewis acids, since they accept a pair of electrons, and the donating species are the Lewis bases, since they donate a pair of electrons. The electron donating species are called *ligands*. These may have a negative charge, such as CN^-, SO_4^{2-}, Cl^-, etc., or they may be neutral species such as H_2O, NH_3, etc. Also, they may be positively charged, $NH_2NH_2^+$, even though these are rare. These transition metal complexes play vital roles in our everyday life. They are useful as important drugs, catalysts, sensors, etc. We know that all these depend on their structures. Hence, it becomes very important to determine their structures. Spectral methods such as electronic (UV-Vis), infrared (IR), nuclear magnetic resonance (NMR), and electron paramagnetic resonance (EPR) play crucial roles in structure determination. These are nondestructive methods and, hence, we need not lose these precious complexes. They are all part of the electromagnetic spectrum and, therefore, we must understand the different regions of the electromagnetic spectrum and the effects they have when they are absorbed by the transition metal complexes. These are briefly described in the following section.

1.2 ELECTROMAGNETIC SPECTRUM

Energy is emitted in the form of electromagnetic radiation. As evident from the name, this consists of two components, namely electrical and magnetic

Spectral Methods in Transition Metal Complexes. http://dx.doi.org/10.1016/B978-0-12-809591-1.00001-3

components. Hence, the radiation is called electromagnetic radiation. It is shown in Fig. 1.1. The frequencies of this radiation range from a very low value of 3 Hz to a very high value of 300 EHz (1 EHz = 10^{18} Hz). The metric prefixes with their symbols are given in Table 1.1.

The electric vector and magnetic vector are mutually perpendicular to the line of propagation. This is shown in Fig. 1.1. This electromagnetic spectrum is divided into various regions based on the frequency.

Table 1.1 Metric Prefixes, Symbols, and Values		
Prefix	Symbol	Value
yotta	Y	10^{24}
zetta	Z	10^{21}
exa	E	10^{18}
peta	P	10^{15}
tera	T	10^{12}
giga	G	10^{9}
mega	M	10^{6}
kilo	k	10^{3}
milli	m	10^{-3}
micro	μ	10^{-6}
nano	n	10^{-9}
pico	p	10^{-12}
femto	f	10^{-15}
atto	a	10^{-18}
zepto	z	10^{-21}
yocto	y	10^{-24}

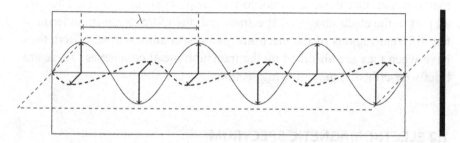

Figure 1.1 Electromagnetic radiation.

1.2.1 Relation Between Frequency and Wavelength

Wavelength

Wavelength is the distance between two crests or two troughs as shown in Fig. 1.1. It is expressed in nm in the "electronic spectrum." 1 nm = 10^{-9} m. Energy of radiation is inversely proportional to its wavelength. That is, when the wavelength increases, energy decreases and when the wavelength decreases, energy increases.

In the IR spectrum, instead of λ or ν, wavenumber $\bar{\nu}$ is used and its unit is cm^{-1}.

The relation between the frequency ν and the wavelength λ is given by Eq. 1.1

$$\nu = \frac{c}{\lambda} \tag{1.1}$$

$$\bar{\nu} = \frac{\nu}{c} = \frac{c}{\lambda} \times \frac{1}{c} = \frac{1}{\lambda} \, cm^{-1} \tag{1.2}$$

where ν is the frequency of radiation, λ is the wavelength of radiation, c is the velocity of light, and $\bar{\nu}$ is the wavenumber.

1.3 REGIONS OF THE ELECTROMAGNETIC SPECTRUM

The electromagnetic spectrum is divided into different regions as shown in Table 1.2.

Table 1.2 Regions of the Electromagnetic Spectrum			
Region	$\bar{\nu}$ (cm^{-1})	λ	ν (Hz)
γ-ray	$>10^8$	>100 pm	$>3 \times 10^{18}$
X-ray	10^6 to 10^8	10 nm to 100 pm	3×10^{16} to 3×10^{18}
UV-Vis	10^4 to 10^6	1 μm to 10 nm	3×10^{14} to 3×10^{16}
Infrared	10^2 to 10^4	100 μm to 1 μm	3×10^{12} to 3×10^{14}
Microwave	1 to 100	1 cm to 100 μm	3×10^{10} to 3×10^{12}
ESR	10^{-12} to 1	100 cm to 1 cm	3×10^8 to 3×10^{10}
NMR	$<10^{-2}$	10 m to 100 cm	3×10^6 to 3×10^8

1.4 EFFECT OF ELECTROMAGNETIC RADIATION ON MATTER

Each type of radiation has a specific interaction with matter, which they strike and produce a unique effect when the radiation is absorbed by the matter [1].

1.4.1 γ-Rays

These have the smallest wavelength and maximum energy. These rays are produced by neutron stars, supernova explosions, regions around black holes, nuclear explosions, lightning, and radioactive decay. The γ-rays will ionize the medium when they pass through it due to their high energy and, hence, are called ionizing radiation. The photons in the γ-rays have 10,000 times more energy than those present in the visible region. These rays have no mass and electrical charges. The γ-rays travel with the speed of light and have the capacity to travel long distances until they lose energy. They cease to exist once their energy is lost. When they fall on a molecule, the configuration of the nucleus changes. That is, *neutron* $\leftrightarrow$ *proton* interconversion takes place, emitting γ radiation. This is shown in Fig. 1.2.

1.4.2 X-Rays

Unlike γ-rays, X-rays are produced artificially. While γ-rays originate from the nucleus, X-rays are produced by electrons surrounding the nucleus. X-rays can be classified as (i) hard and (ii) soft. Hard X-rays have high energy, while the soft X-rays have low energy. When they interact with matter, three things can happen depending on the energy of the X-ray: (i) photoabsorption, (ii) Compton scattering, and (iii) Rayleigh scattering.

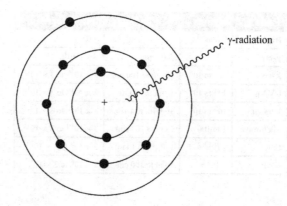

Figure 1.2 Emission of γ radiation from a nucleus.

Photoabsorption takes place when the atomic number of the atom is high and the energy of the X-ray is high. Here, the energy of the X-ray photon is completely transferred to the electron with which it interacts. This causes ionization producing photoelectrons. These photoelectrons may cause further ionization of atoms in their path.

Compton scattering is an inelastic scattering. It is produced when X-ray interacts with soft tissues and thus is useful in medical imaging. This inelastic scattering is produced by an outershell electron. Here, only a portion of the energy from the X-ray is transferred to the electron of interaction causing ionization. This lowers the energy of the X-ray and, hence, the λ of X-ray increases.

Rayleigh scattering is an elastic scattering and is the predominant effect.

In short, when an X-ray interacts with a molecule, it causes a change of electron distribution in the atom(s) of the molecule as shown in Fig. 1.3.

1.4.3 Ultraviolet and Visible (UV-Vis) Radiation
Just like the X-ray, UV-Vis radiation also causes a change in the distribution of electrons. However, the major difference between the two is that an electron from the inner shell of an atom of a molecule (atomic orbital) is excited to a higher shell of the same atom by X-rays, while the UV-Vis radiation excites an electron from the inner molecular orbital to the higher molecular orbital. The effect of UV-Vis on a molecule is shown in Fig. 1.4.

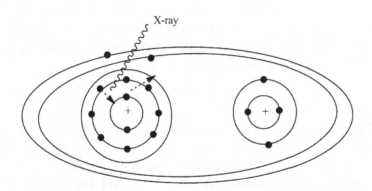

Figure 1.3 Emission of X-ray from a molecule.

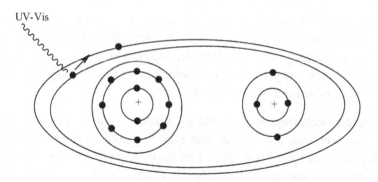

Figure 1.4 Effect of UV-Vis radiation on a molecule.

In other words, UV-Vis radiation causes electronic transition. The electrons can be excited from a σ-orbital to σ^*-orbital known as $\sigma \rightarrow \sigma^*$ transition. This transition requires very high energy and, hence, bands corresponding to this transition will appear in the *far-UV* region (180–200 nm). Therefore, this band will not be visible in the UV region (200–400 nm). This transition appears only in saturated organic molecules, which have only σ-bonds and no π-bonds.

Another possibility is that an electron can be excited from a π-orbital to π^*-orbital, if the ligand has π-bonds, that is, if the molecule is unsaturated. This is known as $\pi \rightarrow \pi^*$ transition. The energy required for this transition falls in the UV region (200–400 nm) and, hence, can be recorded by the UV-Vis spectrophotometer.

The third possibility is that nonbonding electrons can be excited from nonbonding orbitals to either π^* or σ^* orbitals depending on the energy absorbed by a molecule. This is known as $n \rightarrow \pi^*$ transition and $n \rightarrow \sigma^*$ transition. These will be seen in molecules having atoms containing lone pairs of electrons, such as amines (RNH_2), alcohols (ROH), etc., where the "N" and "O" atoms have lone pairs of electrons.

These are shown in Fig. 1.5. However, all of these possibilities provide less information on the structure of the complex as these transitions are related to the ligands. Another type of transition known as a *d-d* transition deals with the transitions of electrons from one *d*-orbital to another. This is responsible for imparting colors to the transition metal ions and their complexes because energy is absorbed in the visible region. These transitions are recorded by the UV-Vis spectrophotometer and provide information on the

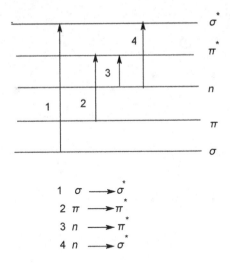

Figure 1.5 Electronic transitions.

structure and geometry of the complex. A detailed discussion of all these factors is provided in the subsequent chapter. Hence, understanding the electronic spectra of complexes is very important.

1.4.4 IR Radiation

When this radiation falls on a molecule and if this is absorbed, then this causes changes in vibration and rotation of the molecule. Hence, the IR spectrum is also known as the "rotation-vibration spectrum." The effect of IR radiation on a molecule is shown in Fig. 1.6. The IR region is from 4000 to 400 cm^{-1} and the far-IR region is from 400 to 200 cm^{-1}. The vibrations can be broadly classified as *stretching* and *bending* vibrations. The stretching vibration is further classified as *symmetric* and *asymmetric* stretching vibrations. Similarly, the bending vibrations are classified as *in-plane* and *out-of-plane* bending vibrations. The energy involved in stretching and bending a bond can be related to the stretching and bending of a spring. As the stretching of a spring requires more energy than bending, stretching of a bond requires more energy than bending a bond. Hence, stretching vibrations appear at a higher wavenumber than bending vibrations. In a similar way, asymmetric stretching vibrations require more energy than the symmetric stretching vibrations.

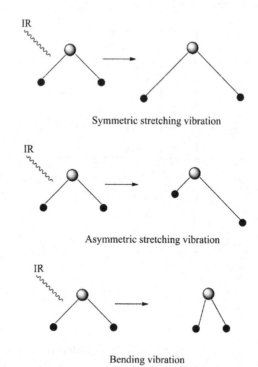

Symmetric stretching vibration

Asymmetric stretching vibration

Bending vibration

Figure 1.6 Effect of IR radiation on a molecule.

Also, the energy required to stretch a bond or bend a bond depends on the reduced mass of the system. Hooke's law is the basis of IR spectra:

$$\nu = \frac{1}{2\pi}\sqrt{\frac{k}{\mu}} \qquad (1.3)$$

where ν is the frequency of vibration, k is the force constant, and μ is the reduced mass of the system. The reduced mass μ is given by

$$\frac{1}{\mu} = \frac{1}{m_A} + \frac{1}{m_B} \quad \text{or} \quad \mu = \frac{m_A m_B}{m_A + m_B} \qquad (1.4)$$

where m_A is the mass of atom A and m_B is the mass of atom B of a molecule AB.

The total number of vibrations for a given molecule is given by the formula $(3N - 5)$ for a linear molecule and $(3N - 6)$ for a nonlinear molecule, where N is the total number of atoms in the molecule, the number 5 and 6 refer to the sum of the translation and rotation for the molecule.

For any molecule, the numbers of translations will be along the three axes and hence will be 3. The rotation will be 2 for a linear molecule and 3 for a nonlinear molecule. For example, consider the molecules CO_2 and H_2O. CO_2 is a linear molecule and, hence, the number of vibrations will be equal to $3 \times 3 - 5 = 4$. Since H_2O is V shaped (ie, a nonlinear molecule), the number of vibrations will be equal to $3 \times 3 - 6 = 3$. These are further discussed in depth in Chapter 3.

1.4.5 Microwave Radiation

Microwave (MW) radiation causes changes in the rotation of molecules, when this radiation falls on the molecule and is absorbed by it. For this reason, "microwave spectrum" is also known as "rotational spectrum." The effect is shown in Fig. 1.7. This is not very useful in the study of complexes. However, it can be used to find the bond length in a molecule.

1.4.6 Nuclear Magnetic Resonance

A spinning nucleus acts as a bar magnet and exhibits precessional motion when placed in an external magnetic field. This precessional frequency (ω) is directly proportional to the applied magnetic field strength (B_0). When ω becomes equal to the radiofrequency (ν) applied, the radiation is absorbed and the nucleus is excited from the ground state to the excited state with spin inversion (flipping). This is called NMR. The spin of a proton, for example, changes from $+1/2$ to $-1/2$. This is shown in Fig. 1.8.

NMR is a general term. When the nucleus is a proton, it is called ^1H NMR or PMR, that is, proton magnetic resonance; when the nucleus is ^{13}C,

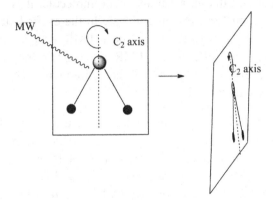

Figure 1.7 Effect of microwaves on a molecule.

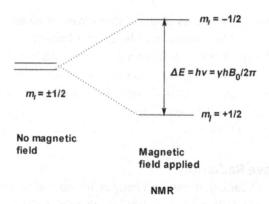

Figure 1.8 Effect of radio frequency on a molecule.

it is called ^{13}C NMR or CMR; when the nucleus is ^{19}F, it is called ^{19}F NMR and when the nucleus exhibiting resonance is ^{31}P, it is called ^{31}P NMR; and so on.

For a nucleus to exhibit NMR, its nuclear spin, $I \neq 0$. NMR has wide application in the structural elucidation of organic molecules. Even though it is not widely used in inorganic chemistry, it is useful in the structural elucidation of some simple *inorganic compounds*, and in the study of metal complexes to find the point of attachment of ligands to metals in cases where there may be ambiguity.

1.4.7 Electron Paramagnetic Resonance

When MW radiation falls on a paramagnetic molecule, it is absorbed and the electron is excited from the ground state to the excited state with spin inversion. The principle of EPR is the same as PMR, except that the ground state has $m_S = -1/2$ in EPR, while in PMR, $m_I = +1/2$. In other words, the order of energy levels is reversed in EPR compared with PMR. The change in electron spin is shown in Fig. 1.9.

EPR is very useful in the study of transition metal complexes. It can be used to find out whether the complex is paramagnetic or not. That is, whether the complex has unpaired electron(s) or not. If the complex contains unpaired electrons, EPR can tell how many unpaired electrons are present in the complex. Also, EPR can help us to find the degree of covalence in the coordinate bond. In addition, we can find from EPR spectrum the orbital in

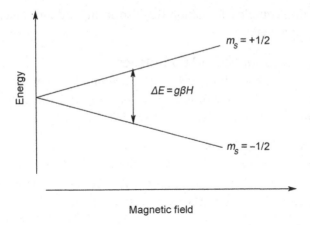

Figure 1.9 Effect of microwave on the spin of electron.

which the unpaired electron is lying. Moreover, EPR spectrum helps us to find whether the complex has undergone any distortion and, if so, what kind of distortion and so on.

1.5 SUMMARY

1.5.1 IR and Complexes

Coordination of a metal to the ligand is confirmed by the IR spectrum of the complex because there will be a change in the position of the band of the functional group coordinating to the metal. Additionally, a new band due to the stretching of the metal-atom will appear. For example, if the ligand contains a carbonyl group, the band due to the stretching frequency of the carbonyl, ν_{CO}, will appear around 1750 cm^{-1}. When the oxygen of the carbonyl group coordinates to the metal to form a complex, ν_{CO} will appear at a lower value than the initial value of 1750 cm^{-1}. Additionally, a new band will appear due to the metal-oxygen stretching in the appropriate region.

In some ligands, there may be more than one coordinating atom. For example, in NCS, both "N" and "S" can coordinate to a metal. This can be easily found out from IR by the appearance of the band corresponding to the metal-atom stretching. If "N" coordinates, the band corresponding to ν_{MN} will appear around 540 cm^{-1} and if "S" coordinates, ν_{MS} will appear in the *far-IR region* around 360 cm^{-1}. Thus, IR is a powerful tool in not only confirming complex formation but also in finding out which atom is coordinating to a metal.

1.5.2 Electronic Spectra and Complexes

Electronic spectra establish the geometry of a complex, that is, octahedral, square planar, etc. The distortion in a complex and the nature of the distortion can also be found from the electronic spectrum.

1.5.3 NMR and Complexes

Whenever there is an ambiguity in the point of attachment of a ligand to a metal, the signals of the protons in the neighborhood of the coordinating atom tell which atom is exactly coordinating to the metal.

1.5.4 EPR and Complexes

This technique tells us whether the complex is paramagnetic or diamagnetic. This technique also gives the degree of ionic character in the coordinate bond. The distortion in the complex and the orbital in which the unpaired electron resides can also be determined from the EPR spectrum.

Thus, the different spectral studies are very helpful in determining the structure of transition metal complexes. In the forthcoming sections each technique will be discussed in detail.

REFERENCE

[1] C.N. Banwell, E.M. McCash, Fundamentals of Molecular Spectroscopy, Tata McGraw Hill, New York, 1994.

CHAPTER 2

Electronic Spectroscopy

The electronic spectrum covers the range from 200 to 800 nm of the electromagnetic spectrum: 200–400 nm is the ultraviolet region and 400–800 nm is the visible region. Hence, the electronic spectrum is also known as the UV-Vis spectrum. The major application of this spectrum to metal complexes is the determination of the geometry of the complex. In order to arrive at this application, we must understand *Term Symbols* first. If we want to understand these concepts, selection rules, and transitions, we must have a clear understanding of the concepts of the symmetry of molecules, the symmetry elements, symmetry operations, point groups of molecules, character tables, direct product concepts, and so on. Hence, we will discuss these concepts before the term symbols.

2.1 SYMMETRY, SYMMETRY ELEMENTS, AND SYMMETRY OPERATIONS

2.1.1 Symmetry

Symmetry means that there will be some uniformity in an arrangement. Many things which we see in our day-to-day life have symmetry and we enjoy them without understanding the concept. Wherever symmetry is present, the appearance will be esthetically pleasing. Some examples are the gates in front of houses, flowers having petals symmetrically arranged (lotus, rose, etc.), high voltage transmission towers, etc.

In the same way, if atoms in a molecule are arranged in some uniform manner, such as six ligands around a central metal atom to give an octahedral complex, two oxygen atoms at equal distance on both sides of a central carbon in CO_2, etc., a symmetrical arrangement results.

2.1.2 Symmetry Elements

How do we distinguish whether a molecule is symmetric or not? We can see that if we are able to find whether the molecule has symmetry elements or not. Then, what are symmetry elements?

Spectral Methods in Transition Metal Complexes. http://dx.doi.org/10.1016/B978-0-12-809591-1.00002-5

These are physical entities such as line, plane, and point, which are not actually present in the molecules but are imaginary. In other words, the symmetry elements can be stated as follows:

1. Identity element, E;
2. Principal or proper axis of symmetry, C_n;
3. Plane of symmetry, σ_v, σ_h, and σ_d;
4. Center of symmetry, i; and
5. Rotation reflection axis or improper axis of symmetry, S_n.

2.1.3 Symmetry Operation

The symmetry operation refers to the action done with the help of the symmetry elements. The symmetry elements and the corresponding operations are given in Table 2.1.

Now the symmetry elements can be defined as follows:

Definition of Symmetry Elements
Principal or Proper Axis of Symmetry, C_n
The proper axis of symmetry or simply axis of symmetry is an imaginary line passing through the molecule about which when the molecule is rotated by a certain angle, $360/n$, an indistinguishable structure results. This is denoted as C_n, where n is called the order of axis. When there is more than one axis of symmetry present in a molecule, the axis with the highest n value is called the *principal axis of symmetry*. Thus, when C_4, C_3, and C_2 axes are present in a molecule, the C_4 axis is called the *principal axis of symmetry*.

C_2 *Axis of Symmetry*   When the molecule is rotated by an angle $360/2 = 180$ degrees, an indistinguishable structure results. For example, a water molecule has a C_2 axis of symmetry, while HOCl has *no* C_2 axis of symmetry as shown in Figs. 2.1 and 2.2.

Table 2.1 Symmetry Elements and Symmetry Operations	
Symmetry Element	**Symmetry Operation**
E	Doing nothing
Proper axis of symmetry	Rotation about the axis
Plane of symmetry	Reflection
Center of symmetry	Inversion
Improper axis of symmetry	Rotation and reflection

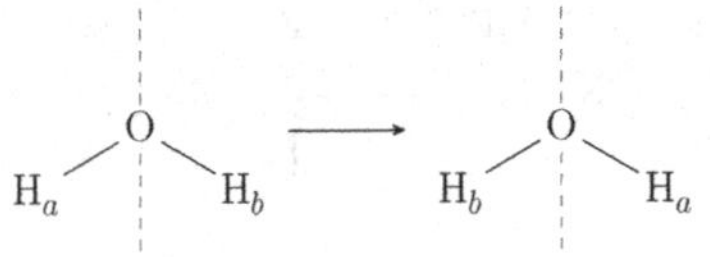

Figure 2.1 Water molecule having C$_2$ axis of symmetry.

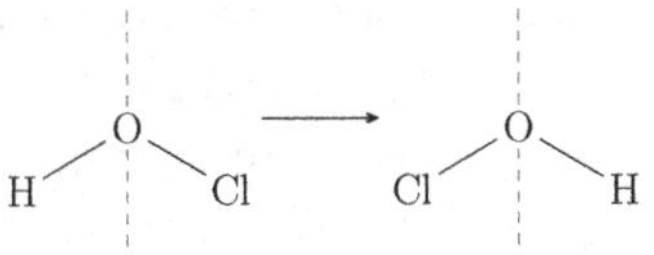

Figure 2.2 Molecule having no C$_2$ axis of symmetry.

When the water molecule is rotated by 180 degrees about the imaginary line passing through 'O', the two hydrogens H_a and H_b are interchanged but the new structure is indistinguishable from the original, when the subscripts 'a' and 'b' are removed (added for our understanding only).

C_3 *Axis of Symmetry*   When a molecule is rotated by 360/3 = 120 degrees by an imaginary line passing through a molecule, if an indistinguishable structure results, the molecule is said to possess a C_3 axis of symmetry as shown in Fig. 2.3. In this molecule the axis is perpendicular to the molecular plane, that is, the plane of paper and passing through 'B' atom. On rotation by 120 degrees about this line counter-clockwise (clockwise is also permitted), F_a, F_b, and F_c are displaced as shown on the RHS and still the new structure will be indistinguishable from the initial one.

C_4 *Axis of Symmetry*   A molecule having a C_4 axis of symmetry will give an indistinguishable structure when rotated by 360/4 = 90 degrees about an imaginary line passing through the molecule as shown in Fig. 2.4. The C_4 axis is perpendicular to the plane of paper and passes through Xe.

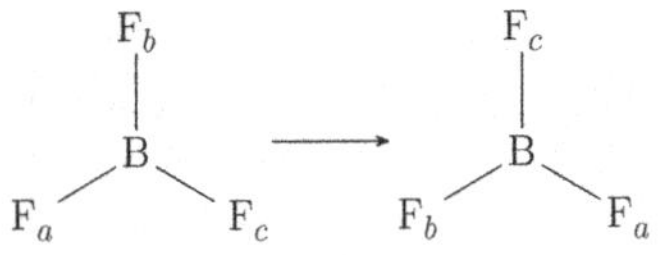

Figure 2.3 Molecule having C$_3$ axis of symmetry.

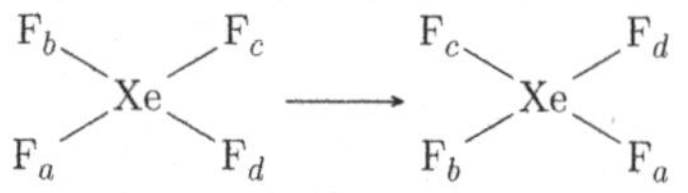

$$\underset{F_a}{\overset{F_b}{\diagdown}}\underset{F_d}{\overset{Xe}{\diagup}}\underset{F_d}{\overset{F_c}{\diagup}} \longrightarrow \underset{F_b}{\overset{F_c}{\diagdown}}\underset{F_a}{\overset{Xe}{\diagup}}\underset{F_a}{\overset{F_d}{\diagup}}$$

Figure 2.4 Molecule having C_4 axis of symmetry.

C_6 *Axis of Symmetry*   When a molecule is rotated by 360/6 = 60 degrees about an imaginary line passing through the molecule, an indistinguishable structure results as shown in Fig. 2.5. In this case, the C_6 axis is perpendicular to the plane of the paper and passes through the center of the molecule.

Plane of Symmetry, σ

The plane of symmetry is an imaginary plane, which cuts the molecule into two halves which are mirror images. It is denoted by the symbol, σ.

Vertical Mirror Plane, σ_v   This is also known as the 'vertical plane of symmetry'. The vertical plane of symmetry is an imaginary plane *coinciding* with the *principal axis* cutting the molecules into two halves which are mirror images as shown in Fig. 2.6 and is denoted by the symbol, σ_v.

Horizontal Mirror Plane, σ_h   The horizontal plane of symmetry σ_h is an imaginary plane perpendicular to the principal axis, reflection on which produces an indistinguishable structure as shown in Fig. 2.7.

Dihedral Plane, σ_d   The dihedral plane of symmetry σ_d is a special type of vertical mirror plane. It bisects two σ_v planes as shown in Fig. 2.8.

Center of Symmetry, i   An imaginary point present at the center of a molecule, about which similar atoms, molecules or groups lie at equal

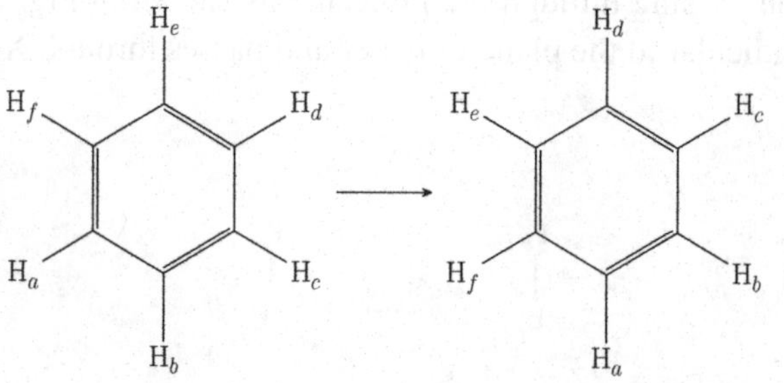

Figure 2.5 Molecule having C_6 axis of symmetry.

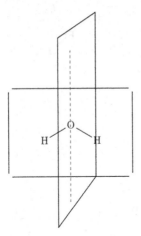

Figure 2.6 Molecule having σ_v planes.

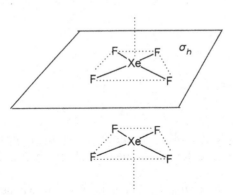

Figure 2.7 Horizontal mirror plane, σ_h.

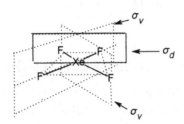

Figure 2.8 Dihedral plane, σ_d.

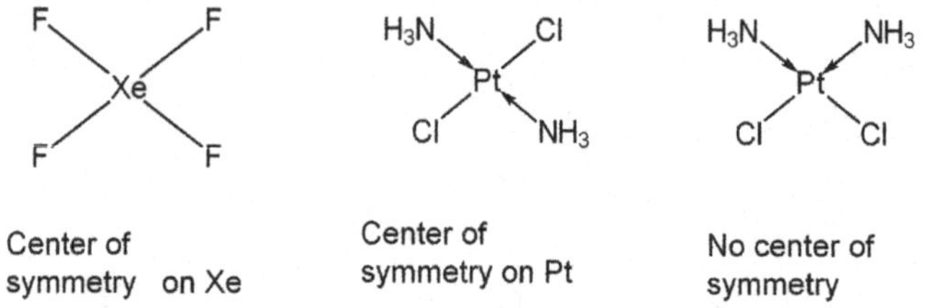

Center of
symmetry   on Xe

Center of
symmetry on Pt

No center of
symmetry

Figure 2.9 Molecule with center of symmetry, i.

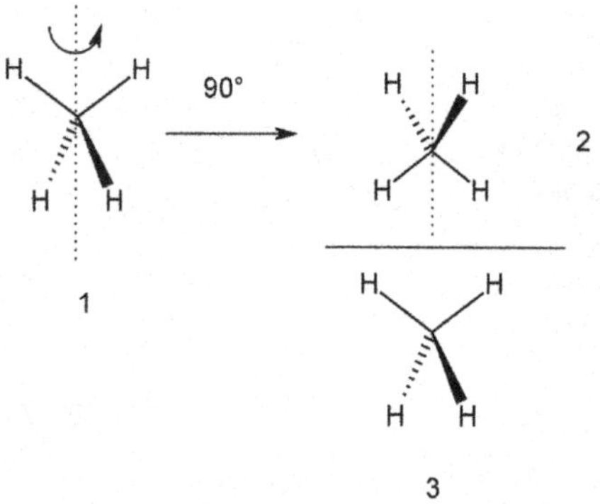

Figure 2.10 Molecule with S₄ axis of rotation.

distances on both sides, is called the center of symmetry or inversion center, C_i. A molecule with a center of symmetry is shown in Fig. 2.9.

Improper Axis of Symmetry, S_n This is also known as the 'Rotation-reflection axis'. When the molecule is rotated by a certain angle, $360/n$ and then reflected in a plane perpendicular to the axis of rotation, if an indistinguishable structure results, the molecule is said to possess an S_n axis of rotation. This is shown in Fig. 2.10.

2.2 IMPORTANT GEOMETRIES OF COMPLEXES

2.2.1 Octahedral Complex

In an octahedral complex, the central metal ion is surrounded by six ligands as shown in Fig. 2.11. There are two positions here, namely, equatorial and axial. The equatorial position is the horizontal square planar arrangement in the xy-plane containing four ligands and the axial positions are the vertical

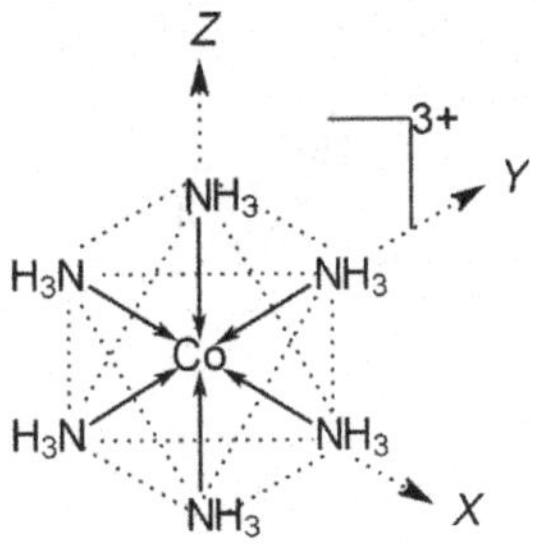

Figure 2.11 Octahedral complex.

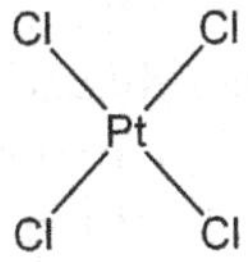

Figure 2.12 Square planar complex.

positions along the z-axis. Here, all the metal-ligand bond angles will be equal and all the bond angles will be equal to 90 degrees. Whenever the bond angles or bond distances vary, the structure will not be a regular octahedron but will be a distorted octahedron. Similarly, even if one ligand is different from the other five, the symmetry will be lowered from O_h.

2.2.2 Square Planar Complex

A square planar complex is considered to be an extremely z-out distorted octahedral complex. That is, when the ligands on both sides of the z-axis are removed to infinity, a square planar complex results. An example for a square planar complex is shown in Fig. 2.12.

2.2.3 Tetrahedral Complex

When four ligands are attached to a central metal ion, either a square planar or tetrahedral complex will result. In the case of a square planar complex, the ligands will be on the x- and y-axes, while in the case of a tetrahedral complex, the four ligands will be approaching the central metal ion in between the axes. An example of a tetrahedral complex is shown in Fig. 2.13.

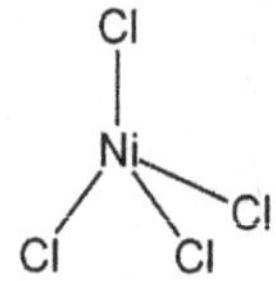

Figure 2.13 Tetrahedral complex.

2.3 TERM SYMBOLS

2.3.1 Terms

Term symbol [1–3] refers to an energy level of a particular arrangement of electrons. While the electronic configuration gives the number of electrons in each orbital, for example, p^2, d^3, etc., term symbols give the different arrangements of electrons in these orbitals, such as, $t_{2g}^3 e_g^1, t_{2g}^2 e_g^2$, etc. Each arrangement of electrons in a shell or orbital has a particular energy. For a p^2 configuration there are 15 different arrangements possible and for a d^2 configuration there are 45 different arrangements possible. Each arrangement has a particular energy. The origin of this energy of a term is the interelectronic repulsion.

There is no interaction between electrons in an unfilled shell and a completely filled shell. For example, in the electronic configuration, $1s^2 2p^2$, there will not be any interaction between the electrons in the s orbital and those in the p orbitals because the s orbital is filled and the p orbital is partially filled. However, there will be interaction between electrons in the partially filled p orbitals.

Electrons in a partially filled orbital of a free-ion exhibit *two* kinds of interactions:

1. Interelectronic repulsion and
2. spin-orbit coupling.

Electron Repulsion Parameters

There are *two* types of interelectronic repulsion parameters:

1. Condon-Shortley parameter (F) and
2. Racah parameter (B and C).

These two parameters are related as follows:

$$B = F_2 - 5F_4 \qquad (2.1)$$

$$C = 35F_4 \qquad (2.2)$$

B is used for the separation between the *same multiplicity* within a configuration, while 'B' and 'C' are used for separation between different multiplicities.

2.3.2 Spin-Orbit Coupling

This phenomenon is the interaction between the magnetic fields produced by the spin and orbital-angular momentum of the electron. This effect is small when compared to the interelectronic repulsion. A term arising from a configuration is split into $(2L + 1)(2S + 1)$ states by the spin-orbit coupling. In other words, a term will be $(2L + 1)(2S + 1)$ degenerate in the absence of spin-orbit coupling.

Types of Spin-Orbit Coupling

The spin-orbit coupling is of *two* types:

1. RS coupling scheme and
2. jj coupling scheme.

RS Coupling Scheme

Russell-Saunders coupling scheme is usually abbreviated as RS coupling scheme. The following points are to be noted for this scheme:

1. It is enough to understand the effect of this on each of the terms and the effect on the other terms are not considered.
2. It is a small perturbation on the terms *individually*.
3. This is valid for *lighter* elements.
4. The RS coupling scheme is not applicable to *heavier* elements because spin-orbit coupling is *stronger* than the interelectronic repulsion.
5. RS coupling splits the configuration of each ion into a number of *states*.
6. States are specified by the total angular momentum quantum number J.
7. The degeneracy of each state is equal to $(2J + 1)$.

jj Coupling

The following points are to be noted with regard to this scheme:

1. In this scheme, spin-orbit coupling is strong and hence important.
2. Configuration is split into *levels* and **not** terms.
3. The levels are specified by spin-orbit coupling.
4. Electron repulsions are treated as perturbations on the spin-orbit coupling levels.
5. This scheme is the direct reverse of the RS coupling scheme.

Spin-Orbit Coupling Parameters

There are *two* parameters, namely ζ (zeta) and λ (lambda).

Parameter ζ

1. It is a single electron parameter.
2. This gives the strength of the interaction between the spin and orbital angular momentum of a single electron of the configuration.
3. This is the property of the configuration.
4. The **operator** corresponding to this perturbation is $\zeta \mathbf{l} \cdot \mathbf{s}$.
5. This parameter is a positive quantity.

Parameter λ

1. This is the property of the term.
2. The operator is $\lambda \mathbf{L} \cdot \mathbf{S}$.

$$\lambda = \pm \frac{\zeta}{2S} \tag{2.3}$$

where the positive sign applies to a shell which is less than half full and the negative sign applies to a shell which is more than half full.

2.3.3 States

States are sub-energy levels arising from terms. These are formed due to the spin-orbit coupling between electrons. That is, the magnetic field produced by the electron due to its spin interacts with that produced by the orbital motion of the electron. It should be remembered that when a charged body rotates or spins, it produces a magnetic field. The following points can be considered:

1. The number of states arising from a term will be equal to $(2S + 1)$ or $(2L + 1)$, whichever is *smaller*.
2. The total angular momentum quantum number J can have values from $|L + S|$ to $|L - S|$.
3. The energy difference between the adjacent states is given by

$$\Delta E_{J,J+1} = \lambda(J + 1) \tag{2.4}$$

4. For shells less than half full, λ is positive and the lowest value of J, namely $|L - S|$ will have the lowest energy.
5. If the shell is more than half full, λ will be positive.

Specification of a State

A state obtained from a term is specified as shown below:

$$^{(2S+1)}L_J$$

The values of L for the different states are the same as those for the different orbitals. Thus, $L = 0$ for S, 1 for P, 2 for D, 3 for F, 4 for G, and so on. When two electrons have parallel spins, $S = 1$ and $2S+1 = 3$ (triplet state). Similarly, when two electrons have antiparallel spins, $S = 0$ and $2S+1 = 1$ (singlet state). A few examples of states are

$$^3P_1, \ ^3F_4, \ ^1G_4$$

Normal and Inverted Multiplets

Splitting of the terms is called *multiplets*. A normal multiplet is one for which the J value increases from the lowest value to the highest value. That is, the lowest level will have the lowest J value. The opposite will be the case of an inverted multiplet. That is, the lowest level will have the highest J value.

Derivation of λ

Let us calculate the λ value for a d^2 system. There are *two* unpaired electrons. Consider Eq. (2.3).

$$S = \frac{1}{2} + \frac{1}{2} = 1 \tag{2.5}$$

$$\lambda = +\zeta/(2 \times 1) \tag{2.6}$$

$$= +\frac{\zeta}{2} \tag{2.7}$$

λ has positive value here because the shell is less than half full (d^5 is half full). Let us consider another example, namely the d^8 system. Here also we have *two* unpaired electrons. However, this is more than half full and hence,

$$\lambda = -\zeta/2$$

2.3.4 Microstates

These microstates originate from states due to the interaction with an external magnetic field. In other words, states are split into microstates in the presence of a magnetic field. In short, we can depict the above as follows:

$$\text{Electronic configuration} \rightarrow \text{Terms} \rightarrow \text{States} \rightarrow \text{Microstates}$$

This means that an electronic configuration is split into terms by inter-electronic repulsion, terms are split into states by spin-orbit interaction, and states are split into microstates by a magnetic field.

2.3.5 Derivation of Term Symbols

A term symbol represents an energy level originating from a configuration. In other words, we can say that it is an abbreviated symbol of energy, angular momentum, and spin-multiplicity of an atom or ion in a particular state. Its derivation for a given electronic configuration involves the following steps:

1. M_L and M_S values are determined.
2. Different electron configurations as allowed by the Pauli principle are determined.
3. A chart of microstates is created.
4. The number of microstates obtained is verified to make sure that all the possible arrangements of electrons have been considered without omitting anything.
5. The chart of microstates is resolved into appropriate atomic states.
6. The ground state of the atom under consideration is determined from among the different states obtained.

Example 1: The different steps can be explained taking the p^2 configuration.

Step 1: Determination of M_L and M_S values.

Determination of Maximum M_L Value
First, we must find the maximum L value for this configuration. As we know, there are three l values for the three p orbitals, namely +1, 0, −1. One of the possibilities is that both the electrons can be present in the p orbital for which $l = +1$. The L value is obtained by multiplying the l value for that orbital by the number of electrons present in that orbital. In this case, $L = 1 \times 2$, which is equal to +2. This is the maximum L value for this p^2 configuration.

Possible M_L Values
After determining the maximum L value, we must find out the allowed M_L values. The possible M_L values, in general, are $L, L-1, L-2, \ldots, 0, \ldots -L$. Thus, in this case, the possible M_L values are 2, 1, 0, −1, −2.

Possible M_S Values

In the given configuration, there are two electrons. The possibilities are that both the electrons can have parallel spins giving rise to $M_S=1$, both of them can have antiparallel spins and $M_S = 0$, or both can have $-1/2$ spin and $M_S = -1$.

Step 2: Determination of electron configurations allowed by the Pauli principle. This can be easily achieved by constructing the pigeonhole diagram as shown in Table 2.2.

Step 3: Chart of microstates creation. Each combination of M_L and M_S is called a microstate. Thus, in the pigeonhole diagram, there are 15 combinations of these values and hence there are 15 microstates. These microstates are arranged in the form of a table. For example, in column 1 of the pigeonhole diagram, $M_L = +1$ and $M_S = +1$. This is entered in the table with an "X" mark. This is shown in Table 2.3.

There is another way of representing the microchart as shown in Table 2.4.

Table 2.2 Pigeonhole Diagram for a p^2 Configuration

M_S	+1	+1	+1	0	0	0	0	0	0	0	0	0	-1	-1	-1
-1	.	↑	↑	.	↓	↓	.	↑	↑	.	.	↑↓	.	↓	↓
0	↑	.	↑	↓	.	↑	↑	.	↓	.	↑↓	.	↓	.	↓
+1	↑	↑	.	↑	↑	.	↓	↓	.	↑↓	.	.	↓	↓	.
M_L	+1	0	-1	+1	0	-1	+1	0	-1	+2	0	-2	+1	0	-1

Table 2.3 Chart of Microstates

.	.	.	M_S	.
.	.	+1	0	-1
.	-2	.	X	.
M_L	-1	X	XX	X
.	0	X	XXX	X
.	+1	X	XX	X
.	+2	.	X	.

Table 2.4 Chart of Microstates—Alternative Method

		M_S		
		+1	0	−1
	+2	.	$[1^+1^-]$	.
	+1	$[1^+0^+]$	$[1^+0^-][1^-0^+]$	$[1^-0^-]$
M_L	0	$[1^+ -1^+]$	$[1^+ -1^-][1^- -1^+][0^+0^-]$	$[1^- -1^-]$
	−1	$[-1^+0^+]$	$[-1^+0^-][-1^-0^+]$	$[-1^-0^-]$
	−2	.	$[1^+1^-]$	.

Step 4: Verification of the number of microstates. The number of microstates that can be obtained from a configuration is given by

$$N = \frac{N_l!}{[x!(N_l - x)!]} \tag{2.8}$$

where N is the total number of microstates,

$$N_l = 2(2l + 1)$$

l is the azimuthal quantum number of the orbital, x is the number of electrons in the orbital. This can be worked out for a p^2 configuration as shown in Eq. (2.12):

$l = 1$ for a p orbital.
$x = 2$ (number of electrons).

$$N_l = 2(2(1) + 1) \tag{2.9}$$
$$= 6 \tag{2.10}$$
$$N = \frac{6!}{[2!(6 - 2)!]} \tag{2.11}$$
$$= 15 \text{ microstates} \tag{2.12}$$

Step 5: Resolving microstates into atomic states.

First, find the maximum value of M_L from the chart of microstates. In this case, it is +2. It represents D state.

$$(0 \to S, 1 \to P, 2 \to D, 3 \to P, 4 \to F, \ldots)$$

An atomic state spans $(2S + 1)$ columns and $(2L + 1)$ rows. An entry corresponding to +2 is present in only one column. Hence, this D state is a singlet state. Thus, it is a ^1D state. Thus, $(2S + 1) = 1$ and therefore this state will span *one* column. This state will span $(2L + 1)$, that is,

$$(2 \times 2 + 1 = 5)$$

rows.

Hence, these can be removed from the chart of microstates and now we are left with Table 2.5.

Now, the maximum M_L value from the chart is equal to +1 and it is a P state. Entries corresponding to +1 are present in *three* columns. Hence, this state is a ^3P state. This spans

$$2L + 1 = (2 \times 1 + 1) = 3$$

rows. These entries can be removed from the microchart and we are left with Table 2.6.

Now, we are left with only one entry with $L = 0$. This refers to S state and this is present in only one column. Thus, it is ^1S. In this way, we have resolved the chart of microstates into the following atomic states:

$$^1D, \ ^3P, \ \text{and} \ ^1S$$

Thus a p^2 configuration gives rise to three terms, namely

$$^1D, \ ^3P, \ \text{and} \ ^1S$$

Table 2.5 Microchart After Removing ^1D

			M_S	
.	.	+1	0	−1
M_L	−1	X	X	X
.	0	X	XX	X
.	+1	X	X	X

Table 2.6 Microchart After Removing ^3P

			M_S	
.	.	+1	0	−1
.	0	.	X.	.

Step 6: Ground state of an atom. The ground state of the different atomic terms obtained in the above manner is determined based on the *three* Hund's rules, which are stated as follows:

Hund's rule 1. The ground state should have the maximum multiplicity. Thus, when there are singlet and triplet terms, the triplet term will be the ground state and, in this case, ^3P will be the ground state.

Hund's rule 2. When two states have the same multiplicity $(2S + 1)$ value, the one with larger L value will have lower energy. Here, we have ^1S and ^1D. Since D has a higher L value than S, ^1D will have lower energy than ^1S. Thus the order of the terms will be

$$^3P < {}^1D < {}^1S$$

However, there may be some violations, since this rule is arbitrary.

Hund's rule 3. This rule gives the relation between J value and the energy of states. The rule is stated as follows: when a subshell is *less than half full*, states with lower J value will be lower in energy. If it is more than half full, then states with higher J value will be lower in energy. In the case of ^3P, the J value is derived as follows:

$$J = L + S$$
$$2S + 1 = 3$$
$$\therefore S = 1$$
$$J = 1 + 1$$
$$= 2$$

J can have values from $L + S, L + S - 1, \ldots, L - S$. Hence, in this case, $J = 2, 1, 0$. Thus, the ^3P will have

$$^3P_2, \, {}^3P_1, \, {}^3P_0$$

states. The p^2 configuration is less than half full. Hence, according to Hund's third rule, the lowest J will have lowest energy. Thus, the three states are arranged as follows according to their energy:

$$^3P_0 < {}^3P_1 < {}^3P_2$$

This is shown in Fig. 2.14.

   Example 2: d^2 configuration.
As in the case of the p^2 configuration, the microchart table for the d^2 configuration can be created and is shown in Table 2.7. As in the case of

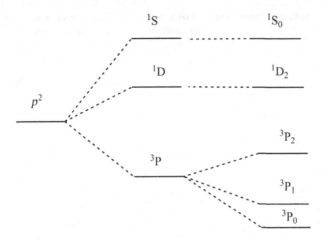

Figure 2.14 The terms arising from a p^2 configuration and the states arising from them.

			M_S		
	4		$[2^+2^-]$		
	3	$[2^+1^+]$	$[2^+1^-][2^-1^+]$		$[2^-1^-]$
M_L	2	$[2^+0^+]$	$[2^+0^-][2^-0^+][1^+1^-]$		$[2^-0^-]$
	1	$[2^+-1^+][1^+0^+]$	$[2^+-1^-][2^--1^+][1^+0^-][1^-0^+]$		$[2^--1^-][1^-0^-]$
	0	$[2^+-2^+][1^+-1^+]$	$[2^+-2^-][2^--2^+][1^+-1^-][1^--1^+][0^+0^-]$		$[2^--2^-][1^--1^-]$

Table 2.7 Microchart for d^2 Configuration

the p^2 configuration, it can be shown that the d^2 configuration gives rise to 45 microstates and they can be resolved to give the following atomic terms:

$$^1G, \, ^3F, \, ^3P, \, ^1D, \, ^1S$$

These terms can be arranged in increasing order of energy based on Hund's rule:

$$^3F < \, ^3P < \, ^1G < \, ^1D < \, ^1S$$

However, while the ground state is ^3F, there may be some violations in the excited states.

Example 3: $4p4d$ splitting. The effect of spin-spin, orbit-orbit, and spin-orbit couplings are shown in Fig. 2.15.

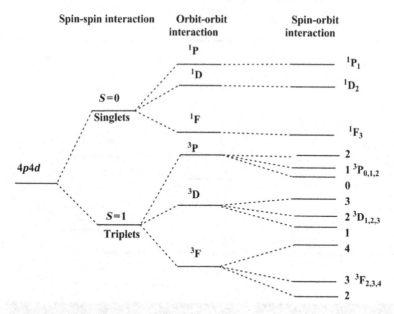

Figure 2.15 *Effect of spin-orbit coupling on pd configuration.*

2.4 SELECTION RULES

The only important mechanism for light absorption by complex ions is the electric dipole mechanism. If absorption is to take place by this mechanism, the transition moment, Q, should be nonzero [4].

Transition Moment Integral, Q

The transition moment integral, Q, can be defined as:

$$Q = \langle \psi_1 | \mathbf{r} | \psi_2 \rangle \tag{2.13}$$

where $\mathbf{r}$ is the radius vector and it has the symmetry of an electric dipole. If the light is polarized, the absorption may be *anisotropic*, which means that the absorption may take place along one direction only. In other directions, there will not be any absorption. If the light is polarized along the z-axis, say, then the transition moment integral will be a function of Q_z. It may be written in the form:

$$Q_z = \langle \psi_1 | z | \psi_2 \rangle \tag{2.14}$$

Components of the Wave Function

This wave function has *five components*, namely *orbital*, *spin*, *vibrational*, *rotational*, and *translational* components. Hence, the wave function may be represented as:

$$\psi = \psi_{\text{orbital}} \cdot \psi_{\text{spin}} \cdot \psi_{\text{vibrational}} \cdot \psi_{\text{rotational}} \cdot \psi_{\text{translational}} \tag{2.15}$$

The last two terms, namely rotational and translational components, will change only in the gas phase and will remain constant in the solid and liquid states; hence, they will be unity on integration in Eq. (2.15). It may be recalled that the transition moment integral, Q, should not be equal to zero in order that a transition will occur (ie, $Q \neq 0$). This is possible only when the conditions given in Eqs. (2.16) and (2.17) are satisfied:

$$\psi_{1,\text{spin}} = \psi_{2,\text{spin}} \tag{2.16}$$

$$\langle \psi_{1,\text{orbital}} \cdot \psi_{1,\text{vibrational}} | \mathbf{r} | \psi_{2,\text{orbital}} \cdot \psi_{2,\text{vibrational}} \rangle \neq 0 \tag{2.17}$$

Spin Selection Rule

This is derived from Eq. (2.16). This equation will be satisfied only when both the wave functions have the same *spin* quantum number, s. This is the *first* selection rule. It can be stated as follows: "The ground and excited states should have the same multiplicity for a transition to take place." This kind of transition is called a multiplicity allowed transition. In other words, singlet ↔ singlet or triplet ↔ triplet transitions are multiplicity allowed transitions.

Laporte Selection Rule

As mentioned earlier, $\mathbf{r}$ is the radius vector and it has the symmetry of an electric dipole. Hence, it is *antisymmetric* to inversion. Therefore,

$$\int_{-\infty}^{+\infty} \psi_{1,\text{orbital}} \cdot \mathbf{r} \cdot \psi_{2,\text{orbital}} = 0, \tag{2.18}$$

if both ψ_1 and ψ_2 have g or u symmetry. Hence, $g \leftrightarrow g$ or $u \leftrightarrow u$ transitions are not allowed. Instead, *$g \leftrightarrow u$ transitions are allowed transitions in molecules having a center of symmetry.* This is the Laporte selection rule. Thus, in *octahedral complexes*, where all the six ligands are equivalent, $d \leftrightarrow d$ transitions are *not* allowed because all the wave functions have g symmetry, when we do not consider vibrations.

2.4.1 Group Theory and Selection Rules

Group theoretical principles help us to find out whether a particular transition is allowed or forbidden [4]. For this, we have to find out whether the transition moment integral, Q, is zero or nonzero. If $Q = 0$ the transition is *not* allowed and if $Q \neq 0$ the transition is allowed.

Direct Product Concept
In order to find out whether $Q = 0$ or $\neq 0$, we must find out whether the reducible representation obtained by the direct product of the irreducible representations of ψ_1, ψ_2, and $\mathbf{r}$ contains the totally symmetric representation of the group A_{1g} or not. If it contains A_{1g}, $Q \neq 0$ and the transition is allowed.

Radius Vector, r   The radius vector, $\mathbf{r}$, transforms as T_{1u} for O_h and as T_2 for T_d.

Example: d^1 octahedral complex. The molecular orbital (MO) diagram for an O_h complex is given in Fig. 2.16.

In total, there will be 13 electrons, 12 from the six ligands and 1 is the metal d electron. The occupancy of the orbitals by electrons is denoted by

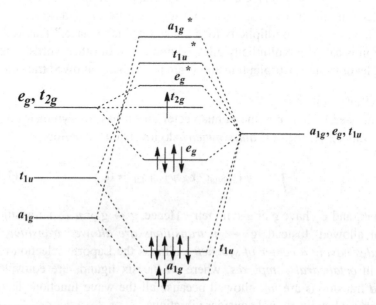

Figure 2.16 MO diagram for O_h complex.

Table 2.8 Character Table for O Group

O	E	$6C_4$	$3C_2(=C_4^2)$	$8C_3$	$6C_2$
A_1	1	1	1	1	1
A_2	1	−1	1	1	−1
E	2	0	2	−1	0
T_1	3	1	−1	0	−1
T_2	3	−1	−1	0	1

arrows in Fig. 2.16. The possible transitions are $t_{2g} \to e_g^*$ and $t_{2g} \to t_{1u}^*$. Of these two transitions, the first one, namely $t_{2g} \to e_g^*$, is forbidden because both the states have 'g' subscripts. Now let us find out from the direct product principle whether the second transition, namely $t_{2g} \to t_{1u}^*$ is allowed or not:

ψ_1 has the symmetry T_{2g}.
ψ_2 has the symmetry T_{1u}.
$\mathbf{r} = x + y + z$ and transforms as T_{1u} in O_h symmetry.
Now the direct product will be $T_{2g} \times T_{1u} \times T_{1u}$.
The O character table is given in Table 2.8.

The irreducible representations of ψ_1, ψ_2, and $\mathbf{r}$ from the character table are written below:

T_2	3	−1	−1	0	1
T_1	3	1	−1	0	−1
T_1	3	1	−1	0	−1
$T_2 \times T_1 \times T_1$	27	−1	−1	0	1

Reduction of the Reducible Representation
Thus, the reducible representation [4] obtained for the direct product is

O	E	$6C_4$	$3C_2(=C_4^2)$	$8C_3$	$6C_2$
Γ	27	−1	−1	0	1

This is reduced using the reduction formula:

$$N = \frac{1}{h}\Sigma \chi_r^x \cdot \chi_i^x \cdot n^x \qquad (2.19)$$

where N is the number of times a particular irreducible representation appears in the reducible representation, h is the total number of operations in the group, which is obtained by adding the coefficients of the symmetry operations in the character table, χ_r^x is the character of the reducible representation, χ_i^x is that of the irreducible representation, and n is the number of operations in the class x and is the coefficient of the operation. Applying the reduction formula, we can reduce the reducible representations as follows:

$$A_1 = \frac{1}{24}[(27 \cdot 1 \cdot 1) + (-1 \cdot 1 \cdot 6) + (-1 \cdot 1 \cdot 3) + (0 \cdot 1 \cdot 8) + (1 \cdot 1 \cdot 6)] = 1$$

$$A_2 = \frac{1}{24}[(27 \cdot 1 \cdot 1) + (-1 \cdot -1 \cdot 6) + (-1 \cdot 1 \cdot 3) + (0 \cdot 1 \cdot 8) + (1 \cdot -1 \cdot 6)] = 1$$

$$E = \frac{1}{24}[(27 \cdot 2 \cdot 1) + (-1 \cdot 0 \cdot 6) + (-1 \cdot 2 \cdot 3) + (0 \cdot -1 \cdot 8) + (1 \cdot 0 \cdot 6)] = 2$$

$$T_1 = \frac{1}{24}[(27 \cdot 3 \cdot 1) + (-1 \cdot 1 \cdot 6) + (-1 \cdot -1 \cdot 3) + (0 \cdot 0 \cdot 8) + (1 \cdot -1 \cdot 6)] = 3$$

$$T_2 = \frac{1}{24}[(27 \cdot 3 \cdot 1) + (-1 \cdot -1 \cdot 6) + (-1 \cdot -1 \cdot 3) + (0 \cdot 0 \cdot 8) + (1 \cdot 1 \cdot 6)] = 4$$

Hence, the reducible representation for the transition $T_{2g} \rightarrow T_{1u}$ contains the following representations:

$$T_{2g} \times T_{1u} \times T_{1u} = A_{1g} + A_{2g} + 2E_g + 3T_{1g} + 4T_{2g}$$

Thus it can be seen that the reducible representation for this transition contains the totally symmetric representation A_{1g} and hence the transition is *allowed*. In this way, the direct product concept of group theory is helpful in predicting whether a given transition will be allowed or forbidden.

Forbidden in O_h but Allowed in T_d The MO diagram for a σ-only tetrahedral complex is shown in Fig. 2.17.

The transition is from e to t_2^*. ψ_1 transforms as E, $\mathbf{r} = x + y + z$ and transforms as t_2 in a tetrahedral field and t_2 transforms as T_2. Hence, the integral is $\langle E \cdot T_2 \cdot T_2 \rangle$. The direct product is obtained from the T_d character table given in Table 2.9.

It can be shown that the direct product is

$$E \times T_2 \times T_2 = 2T_1 + 2T_2 + 2E + A_2 + A_1$$

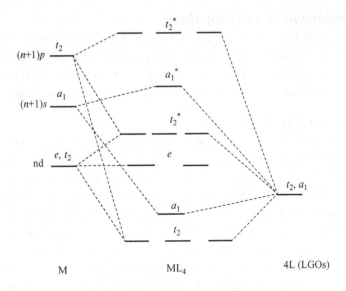

Figure 2.17 MO diagram for tetrahedral sigma only complex.

Table 2.9 T_d Character Table							
T_d	E	$8C_3$	$3C_2$	$6S_4$	$6\sigma_d$	–	–
A_1	1	1	1	1	1	–	$x^2+y^2+z^2$
A_2	1	1	1	−1	−1	–	–
E	2	−1	2	0	0	–	$(2z^2-x^2-y^2, x^2-y^2)$
T_1	3	0	−1	1	−1	(R_x,R_y,R_z)	–
T_2	3	0	−1	−1	1	(x,y,z)	(xy,xz,yz)

This contains A_1 but does not contain 'g'. A_1 alone is not sufficient to find out whether the transition moment integral will be equal to zero or not. The inversion properties must also be foundout, that is, whether 'g' or 'u' is present or not. However, the MOs in T_d do not possess g or u properties because this has no center of symmetry. Even though the MOs do not possess g or u, the atomic orbitals which combine to give the MOs do. The e orbitals are metal d orbitals and they have the g character. The t_2 MOs may contain both p and d orbitals, which have u and d characters, respectively. Hence, the transition moment integral containing the d part will vanish, while that part of the integral containing p part will not. In this way, the forbidden transitions in O_h complexes may be *partially allowed* in T_d complexes due to d-p mixing.

2.4.2 Breakdown of Selection Rules

If a transition is to be seen, then it must obey the selection rules. If this is strictly adhered to, then only very few lines could be seen for the metal complexes in the spectrum. However, more bands are seen than expected. This means that selection rules break down allowing the transitions to take place.

Spin-Selection Rule Breakdown

The spin-selection rule breaks down due to spin-orbit coupling. The intensity of bands will be greater if they are allowed by selection rules and will be weak if they are not. Hence, if the spin-orbital coupling is strong, then the spin-selection rule will break down to a greater extent and the intensity will be greater. Instead, if the spin-orbit coupling is weak, then the spin-selection rule breaks down to a lesser extent and hence the intensity will be weak. The intensity of the spin-forbidden transitions thus obtained in transition metal complexes is always one or two orders of magnitude less than that of the spin-allowed transitions. The spin-orbit coupling magnitude increases as we go from left to right and top to bottom of the periodic table. In other words, the spin-orbit coupling increases as we go from $3d \rightarrow 4d \rightarrow 5d$ transition series.

What Is Spin-Orbit Coupling?   Both orbital motion and spin of an electron produce magnetic fields. The interaction of these two magnetic fields is called spin-orbit coupling and is shown in Fig. 2.18.

Spin-Orbit Coupling of A and E Terms A and E terms do not have any orbital angular momentum because the A term is nondegenerate and contains only one term, while the E term is doubly degenerate because it contains $d_{x^2-y^2}$ and d_{z^2} orbitals, which are shown in Fig. 2.19. The $d_{x^2-y^2}$ orbital cannot be converted into the d_{z^2} orbital by any of the symmetry operations because their shapes are different and hence the orbital angular momentum is absent in these. Hence, the A and E terms do not contribute to spin-orbit coupling.

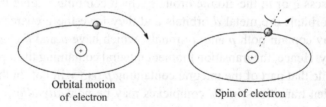

Orbital motion
of electron

Spin of electron

Figure 2.18 Orbital and spin motions of an electron.

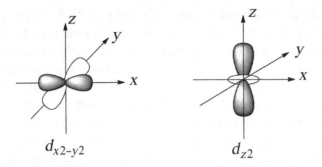

Figure 2.19 The orbitals in the E term.

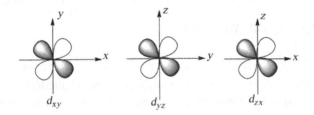

Figure 2.20 The orbitals in the T term.

Spin-Orbit Coupling of the T Term The T term has three d orbitals, namely d_{xy}, d_{yz}, and d_{zx}, which are shown in Fig. 2.20. Since all the three orbitals have similar shape, one can be converted into another by simple rotation. Hence, these orbitals possess orbital angular momentum and contribute to spin-orbit coupling.

Mixing of States and Effect of Spin-Orbit Coupling When the ground state is either A or E, it is not expected to have any orbital angular momentum. However, if the excited state is T and has the same multiplicity as the ground state, then there is mixing of the ground and excited states and, as a result, the ground state acquires some orbital angular momentum and the A and E terms will have some T character. Under these circumstances, the complexes will exhibit significantly higher magnetic moments than the spin-only value, μ_s. For example, when we consider a Ni(II) octahedral complex, its ground state is $^3A_{2g}$. This should exhibit only the normal magnetic moment because the ground term is A. However, this exhibits a higher magnetic moment than expected because this term mixes with the higher T_{2g} term, which has orbital magnetic moment.

Another effect of spin-orbit coupling is that it splits the T terms. As this leads to more energy levels closely spaced, more transitions take place and their energies are very close. This appears as a fine structure in the spectrum. This fine structure can be seen if the spin-orbit coupling is powerful. This fine structure is not observed in solution at room temperature and for the *first* transition metals (ie, $3d$ metals). The reason is that the effect is small. However, in the solid state and at low temperature, the fine structure due to spin-orbit coupling can be seen even in the $3d$ series. In the case of second and third transition metal complexes the fine structure can be seen even in solution because the spin-orbit coupling is large.

The third effect of spin-orbit coupling is the asymmetric shape of the band because the spin-orbit coupling does not split a term symmetrically.

Laporte Selection Rule Breakdown
Vibronic Coupling
This selection rule is applicable only in the case of complexes having a center of symmetry. Hence, even in a symmetric octahedral complex, if the center of symmetry is destroyed by some mechanism such as vibrations, the Laporte selection rule is not obeyed. In other words, the vibrations of the atoms in the ligands in the ground state couple with the excited electronic levels giving rise to vibronic coupling. This removes the center of symmetry in the complex and thus the Laporte selection rule is violated. Due to this vibronic coupling, the otherwise forbidden transition is allowed to some extent. If the ground term is mixed with a g-type vibration, the excited term must be mixed with a u-type vibration and vice versa. Of course these Laporte forbidden transitions are weaker than the allowed transitions.

Intensity Stealing   Intensity stealing is closely related to vibronic coupling. If a forbidden excited term lies near a fully allowed transition, the wave functions of the two terms will mix and, in this process, some intensity of the more intense allowed transition will be taken by the forbidden transition and the net effect is the intensity of the forbidden transition will be greater than expected. This effect decreases when the separation between the allowed and excited terms increases. For example, when some d-d bands lie closer to charge transfer (CT) bands, they are more intense than expected.

Reduction of Symmetry
There is another factor, namely reduction of symmetry, because of which the Laporte selection rule breaks down, that is, violated or relaxed. Consider the transition of a d^2 system in an octahedral complex (O_h), $^3T_{1g} \rightarrow {}^3A_{2g}$.

Table 2.10 Correlation Table for O_h Group

O_h	O	T_d	D_{4h}	D_{2d}	C_{4v}	C_{2v}	D_{3d}	D_3	C_{2h}
A_{1g}	A_1	A_1	A_{1g}	A_1	A_1	A_1	A_{1g}	A_1	A_g
A_{2g}	A_2	A_2	B_{1g}	B_1	B_1	A_2	A_{2g}	A_2	B_g
E_g	E	E	$A_{1g}+B_{1g}$	A_1+B_1	A_1+B_1	A_1+A_2	E_g	E	A_g+B_g
T_{1g}	T_1	T_1	$A_{2g}+E_g$	A_2+E	A_2+E	$A_2+B_1+B_2$	$A_{2g}+E_g$	A_2+E	A_g+2B_g
T_{2g}	T_2	T_2	$B_{2g}+E_g$	B_2+E	B_2+E	$A_1+B_1+B_2$	$A_{1g}+E_g$	A_1+E	$2A_g+B_g$
A_{1u}	A_1	A_2	A_{1u}	B_1	A_2	A_2	A_{1u}	A_1	A_u
A_{2u}	A_2	A_1	B_{1u}	A_1	B_2	A_1	A_{2u}	A_2	B_u
E_u	E	E	$A_{1u}+B_{1u}$	A_1+B_1	A_2+B_2	A_1+A_2	E_u	E	A_u+B_u
T_{1u}	T_1	T_2	$A_{2u}+E_u$	B_2+E	A_1+E	$A_1+B_1+B_2$	$A_{2u}+E_u$	A_2+E	A_u+2B_u
T_{2u}	T_2	T_1	$B_{2u}+E_u$	A_2+E	B_1+E	$A_2+B_1+B_2$	$A_{1u}+E_u$	A_1+E	$2A_u+B_u$

The direct product of ψ_1 and ψ_2 for this transition is: $T_{1g} \times T_{1u} \times A_{2g} = A_{2u} + E_u + T_{2u} + T_{1u}$.

This has no A_{1g} term and hence this transition is *not* allowed in O_h symmetry. Now, if the octahedron is *distorted* by compression or elongation along the z-axis, the symmetry will reduce to D_{4h}. Now, the ground term is no longer T_{1g} but transforms differently. This can be found from the correlation table given in Table 2.10.

From Table 2.9 we can find that T_{1g} in O_h transforms as $A_{2g}+E_g$ in D_{4h}. In the same way, T_{1u} in O_h transforms as $A_{2u} + E_u$ and A_{2g} transforms as B_{1g} in D_{4h}. T_{1u}, which is the dipole moment operator, has *two* components, namely A_{2u} and E_u. A_{2u} is associated with the z-coordinate and E_u is associated with the x- and y-coordinates as a pair. Now, the original direct product has transformed into the new one as follows: $T_{1g} \times T_{1u} \times A_{2g} = (A_{2g}+E_g) \times (A_{2u}+E_u) \times B_{1g}$. The final result can be obtained by referring to the D_{4h} character table shown in Table 2.11.

The final result is that the first term becomes $E_u + B_{1u}$. This has no A_1 symmetry and hence the transition moment integral along the z-coordinate becomes zero and there is no transition along this z-axis. The second term becomes $A_{1u} + A_{2u} + B_{1u} + B_{2u} + E_u$. Since this contains A_1 symmetry, the transition is allowed in the xy-plane (ie, the molecular plane). In other words, the absorption is polarized and the spectrum obtained is an anisotropic and polarized spectrum. Thus the transition is not completely forbidden as in the case of octahedral symmetry but partially allowed when the symmetry

Table 2.11 D_{4h} Character Table

D_{4h}	E	$2C_4$	C_2	$2C_2'$	$2C_2''$	i	$2S_4$	σ_h	$2\sigma_v$	$2\sigma_d$	–	–
A_{1g}	1	1	1	1	1	1	1	1	1	1	–	$x^2 + y^2, z^2$
A_{2g}	1	1	1	−1	−1	1	1	1	−1	−1	R_z	–
B_{1g}	1	−1	1	1	−1	1	−1	1	1	−1	–	$x^2 - y^2$
B_{2g}	1	−1	1	−1	1	1	−1	1	−1	1	–	xy
E_g	2	0	−2	0	0	2	0	−2	0	0	(R_x, R_y)	(xz, yz)
A_{1u}	1	1	1	1	1	−1	−1	−1	−1	−1	–	–
A_{2u}	1	1	1	−1	−1	−1	−1	−1	1	1	z	–
B_{1u}	1	−1	1	1	−1	−1	1	−1	−1	1	–	–
B_{2u}	1	−1	1	−1	1	−1	1	−1	1	−1	–	–
E_u	2	0	−2	0	0	−2	0	2	0	0	(x, y)	–

is reduced to square planar, that is, D_{4h}. The complete working of the direct products is shown below:

$$
\begin{aligned}
T_{1g} \times T_{1u} \times A_{2g} &= (A_{2g} + E_g) \times (A_{2u} + E_u) \times B_{1g} \\
&= [A_{2g}(A_{2u} + E_u) + E_g(A_{2u} + E_u)] \times B_{1g} \\
&= [(A_{2g} \times A_{2u}) + (A_{2g} \times E_u) + (E_g \times A_{2u}) + (E_g \times E_u)] \\
&\quad \times B_{1g} \\
&= [[(A_{1u} + E_u)] + [(E_u) + (A_{1u} + A_{2u} + B_{1u} + B_{2u})]]B_{1g} \\
&= (E_u + B_{1u}) + (A_{1u} + A_{2u} + B_{1u} + B_{2u} + E_u)
\end{aligned}
$$

The first term is concerned with the transition moment integral along the z-direction and the second term is concerned with that along the xy-plane. The former has no term with A_1 symmetry, while the latter has A_1 symmetry. Thus, there is no transition along the z-axis, that is, the C_4 axis of the molecule, while there is a transition along the molecular plane. Hence, the absorption band for this transition is polarized. The transition is still Laporte forbidden because the reduction in symmetry does not alter the inversion properties of the wave functions. It must be overcome by some other mechanism such as vibronic coupling.

2.5 PREDICTION AND ASSIGNMENT OF TRANSITIONS

2.5.1 Orgel Diagram

The possible d-d transitions in a given transition metal complex can be found from the Orgel diagram[1] shown in Figs. 2.21 and 2.22. Two Orgel diagrams, one for the D term and another for the F term are given. We can

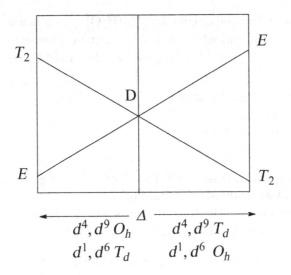

Figure 2.21 Orgel diagram for a D term.

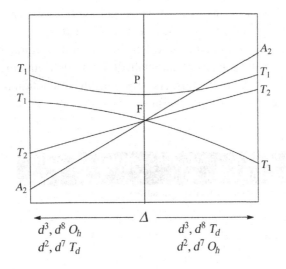

Figure 2.22 Orgel diagram for an F term.

predict the number of possible transitions with the help of these diagrams. For example, from Fig. 2.21, we can predict the transitions for d^1, d^4, d^6, d^9, O_h, and T_d complexes. The corresponding d^n systems and the geometry are labeled on the X-axis. For example, consider the complex, $[Ti(H_2O)_6]^{3+}$. This is a d^1 O_h complex. The ground term is ^2D. This falls on the right of the vertical line in Fig. 2.21. Hence, the transition will be $^2T_{2g} \rightarrow {}^2E_g$.

Similarly, if we take another complex, $[Cr(H_2O)_6]^{3+}$, it is a d^3 octahedral complex. We must see the left-hand side to the vertical line in Fig. 2.22. The transitions will be $^4A_{2g} \rightarrow {}^4T_{2g}$, $^4A_{2g} \rightarrow {}^4T_{1g}(F)$, and $^4A_{2g} \rightarrow {}^4T_{1g}(P)$.

Limitations of the Orgel Diagram
1. It is only qualitative.
2. It applies to only weak-field cases.

2.5.2 Configuration, Free-Ion Term, Ground, and Excited Terms in a Weak Octahedral Field

The d^n configurations, the free-ion terms arising from them, and the spectroscopic terms due to ligand field splitting are given in Table 2.12.

It is clear from Table 2.12 that the ground term can be either D or F in a weak-field case. Thus, it becomes necessary to understand how these *free-ion terms* are split by ligands in order to predict the transitions. The splitting of these two terms arising from d^1 and d^2 is shown in Fig. 2.23A and B, respectively. These are the terms in spherical symmetry. When we move from spherical to octahedral symmetry, there is a reduction of symmetry because in spherical symmetry there will be an infinite number of ligands surrounding the metal ion, while in octahedral symmetry the metal ion will be surrounded by only six ligands. As a result of this reduction in symmetry, the terms will be split as shown in Table 2.13.

d^n	Free-Ion Term	Ground Term	Excited Terms
d^0	^1S	.	.
d^1	^2D	$^2T_{2g}$	^2E_g
d^2	^3F	$^3T_{1g}(F)$	$^3T_{2g}, {}^3A_{2g}, {}^3T_{1g}(P)$
d^3	^4F	$^4A_{2g}$	$^4T_{2g}, {}^4T_{1g}(F), {}^4T_{1g}(P)$
d^4	^5D	^5E_g	$^5T_{2g}$
d^5	^6S	$^6A_{1g}$	
d^6	^5D	$^5T_{2g}$	^5E_g
d^7	^4F	$^4T_{1g}(F)$	$^4T_{2g}, {}^4A_{2g}, {}^4T_{1g}(P)$
d^8	^3F	$^3A_{2g}$	$^3T_{2g}, {}^3A_{1g}(F), {}^3T_{1g}(P)$
d^9	^2D	^2E_g	$^2T_{2g}$
d^{10}	^1S		.

Table 2.12 Configurations, Free-Ion Terms, and Spectroscopic Terms

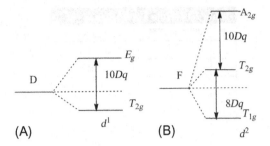

Figure 2.23 Splitting of D and F terms.

Table 2.13 Splitting of Terms in an O_h Environment		
Term	Degeneracy	States in O_h Field
S	1	A_{1g}
P	3	T_{1g}
D	5	$E_g + T_{2g}$
F	7	$A_{2g} + T_{1g} + T_{2g}$
G	9	$A_{1g} + E_g + T_{1g} + T_{2g}$
H	11	$E_g + T_{1g} + T_{1g} + T_{2g}$
I	13	$A_{1g} + A_{2g} + E_g + T_{1g} + T_{2g} + T_{2g}$

The possible transitions in all the d^n cases can be predicted with the help of these two diagrams alone. The basis for this is the *hole formalism*.

Hole Formalism

A "hole" is nothing but a vacancy in the d^n system. Thus, a d^9 system can be considered as a d^{10} system with *one* hole and a d^8 system can be considered to be a d^{10} system with *two* holes.

Advantage of Hole Formalism

The advantage of this *hole formalism* is that a d^9 system can be considered to be an inverted d^1 system and a d^8 system can be considered to be an inverted d^2 system. Hence, the transitions in the d^9 and d^8 systems can be obtained from inverted Fig. 2.23A and B, respectively. In the same way, other systems can be expressed in terms of these two figures. However, the multiplicity has to be considered. These are given in Table 2.14.

Table 2.14 Hole Formalism		
d^n	d^{10-n} or $d^{5\pm n}$	Figure
d^1	–	Fig. 2.10A
d^2	–	Fig. 2.10B
d^3	d^{5-2}	Inverted Fig. 2.10B
d^4	d^{5-1}	Inverted Fig. 2.10A
d^5	–	–
d^6	d^{5+1}	Fig. 2.10A
d^7	d^{5+2}	Fig. 2.10B
d^8	d^{10-2}	Inverted Fig. 2.10B
d^9	d^{10-1}	Inverted Fig. 2.10A

Now, it will be easy to understand why we have Orgel diagrams for D and F terms only. *It should not be forgotten that all the above discussions hold good for weak-field cases only.*

2.5.3 Splitting of d-Orbitals in Different Geometries

The splitting of d-orbitals in octahedral, tetrahedral, trigonal planar, square planar, trigonal bipyramid, and square pyramid geometries are given in Fig. 2.24A-F, respectively.

2.5.4 Free Ions in Medium and Strong Crystal Fields

Strong-Field Configurations

It is very important to understand how the energy levels arising from a free ion are named when they are surrounded by ligands (i.e., crystal fields). We need to find how these levels transform themselves in the *octahedral* (O_h) and *tetrahedral* (T_d) symmetries. Once the naming for a level is derived for O_h symmetry, its name in T_d symmetry can be easily obtained by dropping the subscript 'g'. Thus, if the name for a level in O_h symmetry is A_{2g}, its name in T_d symmetry will be A_2.

In a strong field, the interelectronic repulsions are small compared to those in the crystal field; that is, the repulsion between the ligands and the central metal ion. Hence, the free-ion terms are split to a greater extent compared to the separation between them.

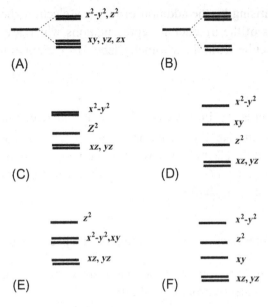

Figure 2.24 Splitting of d-orbitals under various geometries. (A) Octahedral. (B) Tetrahedral. (C) Trigonal planar (ligands lie in xy plane). (D) Square planar. (E) Trigonal bipyramid (pyramid in xy plane). (F) Square pyramid (pyramid in xy plane).

2.5.5 Weak to Strong-Field Transition

In the previous case, the crystal field was stronger than the interelectronic repulsion. However, now both will be of comparable magnitude. In other words, at medium field strength, the splitting of the terms and the separations between them will be more or less equal.

Free-Ion Terms and Cubic Terms

When interelectronic repulsions split the free-ion configurations into terms, they are called free-ion terms. However, when crystal fields in octahedral or tetrahedral environments split these terms, they are called *cubic terms*.

2.5.6 Terms Arising From a Strong-Field Configuration

The cubic terms which can originate from strong-field configurations can be determined by knowing the direct product of the irreducible representations of the initial and final stages. Let us assume that we know the symmetry label of the initial term from a configuration. Let us add one electron to this configuration. A new term will arise now. We know the symmetry label of the added electron. Let us find out the direct product between the symmetry labels of the initial term and that of the added electron. The rule is that the new terms arising by the addition of the new electron should have the symmetry labels of the irreducible representations which are contained in the direct product between the symmetry labels of the initial terms and the added electron.

d^1 System

The configuration is t_{2g}^1. In this case, there is no interelectronic repulsion. Hence, the symmetry label of the terms arising from the d^1 configuration is the same as that of the electron. The wave functions of the term and electron are identical. Thus, in an O_h field, a t_{2g} electron gives rise to $^2T_{2g}$ and an e_g electron gives rise to ^2E_g terms.

d^2 System

Now, let us add one electron and the configuration becomes t_{2g}^2.

t_{2g}^2 *System*   Let us carry out the following steps to arrive at the symmetry labels of the terms arising from this addition.

1. The symmetry label of the original term of the configuration t_{2g}^1 is T_{2g}.
2. The symmetry label of the added electron is T_{2g}.
3. The direct product of the symmetry labels is $T_{2g} \times T_{2g}$. This is obtained from the rules given in Table 2.15. The result is: $T_{2g} \times T_{2g} = A_{1g} + E_g + T_{1g} + T_{2g}$. That is, the t_{2g}^2 configuration gives rise to four terms having the symmetries $A_{1g}, E_g, T_{1g}, T_{2g}$.

Special results are shown in Table 2.16.

The complete results for O are given in Table 2.17.

$t_{2g}^1 e_g^1$ *System*   This is an excited configuration of d^2. The direct product is $T_{2g} \times E_g$. It can be found out from the table that $T_{2g} \times E_g = T_{1g} + T_{2g}$.

Case: e_g^2. This is the second excited state from the configuration d^2. The direct product is $E_g \times E_g$ and the result from Table 2.16 is $E_g \times E_g = E_g + A_{1g} + A_{2g}$. After determining the symmetry labels of the various terms

Table 2.15 Rules for Evaluation of Direct Products in Symmetry Groups

$A \times A = B$	$B \times A = B$	$E \times A = E$	$T \times A = T$	$g \times g = g$	$' \times ' = '$	$1 \times 1 = 1$
$A \times B = B$	$B \times B = A$	$E \times B = E$	$T \times B = T$	$g \times u = u$	$' \times '' = ''$	$1 \times 2 = 2$
$A \times E = E$	$B \times E = E$	$E \times E = *$	$T \times E = T_1 + T_2$	$u \times g = u$	$'' \times ' = ''$	$2 \times 1 = 2$
$A \times T = T$	$B \times T = T$	$E \times T = T_1 + T_2$	$T \times T = *$	$u \times u = g$	$'' \times '' = '$	$2 \times 2 = 1$

Table 2.16 Special Results

For E:

In many groups, including O, T_d, C_{3v}, and D_6:

$E_1 \times E_1 = E_2 \times E_2 = A_1 + A_2 + E_2$

$E_1 \times E_2 = E_2 \times E_1 = B_1 + B_2 + E_1$

If there are **no subscripts** to one of A, B, or E, read $A_1 = A_2 = A$, etc.

$E \times E = A_1 + A_2 + B_1 + B_2$

For T:

$T_1 \times T_1 = T_2 \times T_2 = A_1 + E + T_1 + T_2$

$T_1 \times T_2 = T_2 \times T_1 = A_2 + E + T_1 + T_2$

Examples:

In O_h:

$E_g \times T_{1u} = T_{1u} + T_{2u}$

$A_{2u} \times T_{2u} = T_{1g}$

$E_g \times E_g = A_{1g} + A_{2g} + E_g$

In D_{4h}:

$E_g \times B_{2g} = E_g$

$E_g \times E_u = A_{1u} + A_{2u} + B_{1u} + B_{2u}$

In C_{3v}:

$A_1 \times A_2 = A_2$

$A_2 \times E = E$

$E \times E = A_1 + A_2 + E$

In D_{3h}:

$E' \times A_2' = E'$

$E'' \times E' = A_1'' + A_2'' + E''$

Table 2.17 Complete Results for O

O	A_1	A_2	E	T_1	T_2
A_1	A_1	A_2	E	T_1	T_2
A_2	A_2	A_1	E	T_2	T_1
E	E	E	$A_1 + A_2 + E$	$T_1 + T_2$	$T_1 + T_2$
T_1	T_1	T_2	$T_1 + T_2$	$A_1 + E + T_1 + T_2$	$A_2 + E + T_1 + T_2$
T_2	T_2	T_1	$T_1 + T_2$	$A_2 + E + T_1 + T_2$	$A_1 + E + T_1 + T_2$

arising from a configuration, their multiplicities are found as follows: The three t_{2g} orbitals are represented by a set of three boxes and the two electrons can be arranged in 15 different ways. [The three boxes may be considered as three p orbitals. Then $(2L + 1)(2S + 1) = 5 \times 3 = 15$ because $L = 2$ for D and $S = 1$ is the maximum value for the two electrons. Thus, we have 15 wave functions for the four terms originating from t_{2g}^2.] Let the multiplicities of the four terms be a, b, c, and d, respectively. That is, the four terms can be represented as follows: $^aT_{2g} + ^bT_{1g} + ^cE_g + ^dA_{1g}$. The multiplicities can either be 1 or 3 from the two electrons. Hence, a, b, c, d can have values either 1 or 3.

Thus, we should have

$$3a + 3b + 2c + d = 15, \qquad (2.20)$$

by putting the proper degeneracies as coefficients ($T = 3$, $E = 2$, and $A = 1$). The values for a, b, c, and d are found by a trial and error method as follows:

Case (i): $a = b = c = d = 1$.
  When we put $a = b = c = d = 1$ in Eq. (2.20), we get
  $3(1) + 3(1) + 2(1) + 1 = 9$ and the equation is not satisfied.
Case (ii): $a = b = c = d = 3$.
  When we put $a = b = c = d = 3$, we get
  $3(3) + 3(3) + 2(3) + 3 = 27$ and the equation is again not satisfied.
  Let us try with a mixture of 1 and 3.
Case (iii): $a = 1$, $b = 1$, $c = 1$, $d = 3$.
  We get $3(1) + 3(1) + 2(1) + 3 = 11$ and equation is not satisfied.
Case (iv): $a = 1$, $b = 1$, $c = 3$, $d = 1$.
  Let us put $a = 1$, $b = 1$, $c = 3$, and $d = 1$. Then
  $3(1) + 3(1) + 2(3) + 1 = 13$ and the equation is not satisfied.
Case (v): $a = 1$, $b = 3$, $c = 1$, $d = 1$. The other possibility is: $a = 1$,
  $b = 3$, $c = 1$, $d = 1$. Now we get $3(1) + 3(3) + 2(1) + 1 = 15$ and the
  equation is satisfied.
Case (vi): $a = 3$, $b = 1$, $c = 1$, $d = 1$.
  By putting in these values, we get $3(3) + 3(1) + 2(1) + 1 = 15$.
  In short, one of the T terms should be a triplet, while other terms are
  singlets. It can be finally shown that the T_{1g} term is a triplet. Thus,
  $t_{2g}^2 \rightarrow {}^3T_{1g} + {}^1T_{2g} + {}^1E_g + {}^1A_{1g}$.

Lowest Term of a Strong-Field Configuration

Hund's rule tells us that the term with highest orbital degeneracy will be the ground state. Thus one of the T terms should be the ground term. An additional Hund's rule says that that term with maximum multiplicity will have the lower energy. Hence, in this case, $^3T_{1g}$ is the ground term from a d^2 configuration.

e_g^2 *Configuration*   This is another excited state of d^2 configuration, where both the electrons are in the e_g orbital. The symmetries of the terms arising from this configuration are obtained from Tables 2.16 and 2.17. They are $A_{1g} + A_{2g} + E_{2g}$.

d^3 Strong-Field Configuration
The possible configurations are t_{2g}^3, $t_{2g}^2 e_g^1$, $t_{2g}^1 e_g^2$, and e_g^3. When the t_{2g} and e_g levels are more than half full, the *hole formalism* operates. That is, e_g^3 is equivalent to e_g ($e_g^3 \equiv e_g$).

Strong and Weak-Field Cases: Difference
Weak-Field Case   Here, interelectronic repulsion is more powerful than the crystal field effect.

Strong-Field Case   Here, the crystal field is more powerful than interelectronic repulsion.

Noncrossing Rule
This rule is to be considered when we follow the transition from weak field to strong field. According to this rule, the wave functions which have the *same symmetry* will *never cross*, when the energy is plotted against the perturbing potential. For example, $^3T_{1g}$ comes from a P term and an F term. Their wave functions will never cross as shown in Fig. 2.9.

2.5.7 Tanabe-Sugano Diagrams
These diagrams are drawn taking energy of the terms on the Y-axis and Δ_o on the X-axis [1]. In fact, these are divided by the Racah parameter, B, and then plotted. In other words, E/B is taken on the Y-axis and Δ_o/B on the X-axis. Each d^n system has a separate diagram. Unlike Orgel diagrams, these are applicable to strong-field cases also. Whenever a vertical line is there in the diagram, the left-hand side of the vertical line refers to the weak-field case and the right-hand side applies to the strong-field cases. These give quantitative information also. Δ_o and B can be calculated from the diagram, once the energies of transitions are known from the spectrum of a complex. For example, let us consider a d^2 system, whose Tanabe-Sugano diagram is partially given in Fig. 2.25. Only the important terms are shown in the figure. The important atomic terms arising from a d^2 configuration are $^3F < {}^1D < {}^3P$. The ^3F atomic term splits into $^3T_{1g}$ (x-axis), $^3T_{2g}(F)$ and $^3A_{2g}(F)$ spectroscopic terms. The first excited atomic term, ^1D splits into ^1E_g and $^1T_{2g}$ spectroscopic terms. The second excited atomic term, ^3P

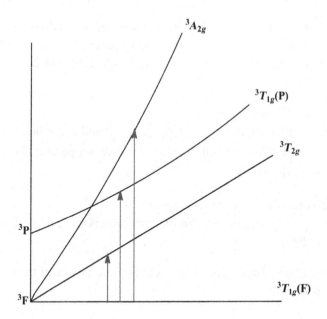

Figure 2.25 Tanabe-Sugano diagram for a d^2 complex.

is not split but transformed into the $^3T_{1g}(P)$ spectroscopic term. These are shown in Fig. 2.25. Then the allowed transitions are assigned as

$$^3T_{1g}(F) \to {}^3T_{2g}(F), \quad {}^3T_{1g}(F) \to {}^3T_{1g}(P), \quad {}^3T_{1g}(F) \to {}^3A_{2g}(F)$$

and are shown by three arrows from the x-axis because the ground term coincides with the x-axis. The energies of these transitions depend upon the ligand field strength, that is, on the Δ_o value. Hence, if the energy of a particular transition is very high, it may not fall in the visible region and hence will not be seen in the spectrum. In this way, the number of transitions can be predicted from the respective Tanabe-Sugano diagram and assigned.

There is no question of high-spin or low-spin cases until the d^3 system and there will be only one case, namely the high-spin case. However, from d^4 onward, the strong-field case also comes into play. Hence, we can have both low-spin (strong-field) and high-spin (weak-field) configurations. For example, let us consider the d^4 system. Its partial Tanabe-Sugano diagram is given in Fig. 2.26.

If the ligand is a weak ligand, we must see the left-hand side of the vertical line in Fig. 2.13. Now the ground term is ^5E_g and coincides with the x-axis. The excited state is $^5T_{2g}$. Hence, there is only one spin-allowed

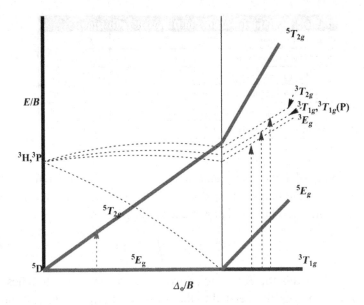

Figure 2.26 Tanabe-Sugano diagram for d^4 system.

transition, namely $^5E_g \rightarrow {}^5T_{2g}$. On the other hand, if the ligand is strong, we must see the right-hand side of the vertical line in Fig. 2.26. The ground state is $^3T_{1g}$ coinciding with the x-axis. The excited states with the same multiplicity are ^3E_g, $^3T_{2g}$, $^3A_{1g}$, and $^3A_{2g}$. The transitions to these states are spin allowed. However, depending on the energy of transitions, we may be able to see them in the visible region. Thus at least the transitions $^3T_{1g} \rightarrow {}^3E_g$ and $^3T_{1g} \rightarrow {}^3T_{2g}$ may be seen in the visible region of the electronic spectrum. In short, if the ligand is weak we will have one transition and if the ligand is strong we will have a minimum of two spin-allowed transitions. In this manner, the Tanabe-Sugano diagrams will be helpful in predicting and assigning the transitions. Later, we will see how these Tanabe-Sugano diagrams will give quantitative information.

2.6 BAND INTENSITIES, BAND WIDTHS, AND BAND SHAPES

2.6.1 Band Intensities

The band intensity is expressed in terms of extinction coefficients, ϵ, assuming that the band widths are equal. However, band widths are temperature dependent. Hence, we must be careful in comparing band intensities in terms of ϵ. It will be reasonable to compare at the same temperature. The intensities of different kinds of transitions are summarized in Table 2.18.

Table 2.18 Band Intensities of Different Transitions		
Type of Transition	f	ϵ
Spin-forbidden, Laporte forbidden	10^{-7}	10^{-1}
Spin-allowed, Laporte forbidden	10^{-5}	10
Spin-allowed, Laporte forbidden but with d-p mixing (T_d)	10^{-3}	10^2
Spin-allowed, Laporte forbidden but with intensity stealing	10^{-2}	10^3
Spin-allowed, Laporte allowed (CT)	10^{-1}	10^4

The intensity of a band is measured in terms of oscillator strength, f. This is the area under the band obtained when extinction coefficient ϵ is plotted against frequency ν. The oscillator strength f is given by

$$f = 4.32 \times 10^{-9} \int_{\nu_2}^{\nu_1} \epsilon \, d\nu \tag{2.21}$$

where ν_1 and ν_2 are the frequencies over which the band stretches. The oscillator strength f is also related to the transition moment Q according to

$$f = 1.096 \times 10^{11} \nu Q^2 \tag{2.22}$$

One assumption is made that both the ground and excited states are nondegenerate.

2.6.2 Band Widths
In electronic spectra, the band widths are wide because the terms are spread over a range of energies. That is, the terms have in themselves a large number of closely spaced energy levels. As a result, a large number of closely spaced transitions will take place and these will merge to give a broad band. Several factors may play a role in deciding the shape of a band, such as

- vibration;
- spin-orbit coupling; and
- Jahn-Teller effect.

Vibration and Band Width
The splitting of d-orbitals or 10Dq or Δ_o depends not only on the strength of the ligand but also on the metal-ligand separation because when the ligands are closer to metals, the repulsion between the metal and ligand electrons will be greater and the splitting will be increased. Thus, the distance between the metal and ligand varies due to vibration and hence the energy of each term is spread over a range of values. This increases the band width. If the

energies of two terms depend on Δ_o but in different manners, then the energies of these terms will be spread over a range giving rise to a broad band. Sometimes the band width may be as much as 1000 cm^{-1}. However, if the ground and excited terms are parallel at least partially, and if the energies of these two terms do not depend on Δ_o, then these transitions will not be broadened due to a vibrational mechanism.

Spin-Orbit Coupling and Band Width
Spin-orbit coupling splits the T terms and more closely spaced energy levels are formed. Thus, more transitions than expected will be observed and these will be dependent on the magnitude of spin-orbit coupling and the broadening will be from 100 to 1000 cm^{-1}.

Jahn-Teller Effect and Band Width
Jahn-Teller effect operates when the ground state is degenerate and this effect splits the degenerate levels causing broadening of bands. The broadening may be as much as 1000 cm^{-1}.

CT Bands and Band Width
The CT is even broader than 1000 cm^{-1}. The reason is that there are several excited terms and transitions to all these terms will take place from the ground term. All these transitions will be very close and hence merge to give a broad band.

2.6.3 Band Shapes
The factors determining the band shapes are given below:

1. vibrational interaction
2. spin-orbit coupling and
3. low-symmetry ligand fields.

Vibrational Interaction and Band Shape
The bands are quite often asymmetric instead of symmetric and this is due to the vibrational interaction. If the intensity of a transition is largely due to vibronic coupling, then a symmetric band can be expected only at low temperatures. As the temperature increases, the shape becomes asymmetric and a tail starts appearing toward the low frequency. Also, the band maximum may not give the exact energy difference between the ground and excited terms.

Spin-Orbit Coupling and Band Shape

Another reason for the asymmetry of band shape is the spin-orbit coupling. Spin-orbit coupling removes the degeneracies of terms, if present. That is, spin-orbit coupling causes splitting of terms, but this splitting is not uniform and hence the band due to this transition is not symmetric in shape.

Lowering of Symmetry

When all the ligands are equivalent, the symmetry of the complex will be high. For example, the complex $[Co(NH_3)]_6^{3+}$ has high symmetry, namely O_h, because all the six ligands are the same. Instead, even if one NH_3 is replaced by another ligand, Cl^- for example, the symmetry is no longer O_h but is lowered. This lowering of symmetry in turn splits the E and T terms. However, the splitting pattern differs. While E term is split into two symmetrically, T term is asymmetrically split into one component of twofold degeneracy and another nondegenerate component. Hence, the band arising due to transition between these terms will be asymmetric in shape.

In spite of these factors, spectra obtained in solution are nearly symmetric. The reason may be that all these three factors may simultaneously operate to give an averaged shape, which may be nearly symmetric.

2.7 COMPLEXES AND COLOR

The transition metal complexes exhibit different colors due to d-d transitions. If a complex appears green, this is not the original color of the complex. This is the color emitted by the complex. Actually, the complex absorbs orange color and emits its complementary color, green. The color absorbed and the corresponding complementary color are given in Table 2.19 and the color and the corresponding energies are given in Table 2.20.

2.8 ELECTRONIC SPECTRA OF INDIVIDUAL IONS

2.8.1 Electronic Spectrum of a d^1 System

The electronic spectrum [5, 6] of aqueous solution of Ti^{3+} is shown in Fig. 2.27. The aqueous solution is purple in color. That is, this is the color emitted by the complex. From Table 2.19, we can find out that the color absorbed will be green. From Table 2.20, it can be seen that the approximate energy of this color will be 20,000 cm^{-1}. This is confirmed by the absorption around 20,000 cm^{-1} in its electronic spectrum shown in Fig. 2.27. The transition has been assigned as $^2T_{2g} \rightarrow {}^2E_g$.

Table 2.19 Color and Complementary Color	
Color Absorbed	**Color Transmitted**
Violet	Yellow-green
Blue	Yellow
Green-blue	Orange-red
Green	Purple
Yellow-green	Violet
Yellow	Blue
Orange	Green-blue
Red	Blue-green

Table 2.20 Colors and Their Energies	
Color	**App. Energy (cm^{-1})**
Red	14,000–16,000
Yellow	18,000
Green	20,000
Blue	21,000–25,000
Violet	25,000

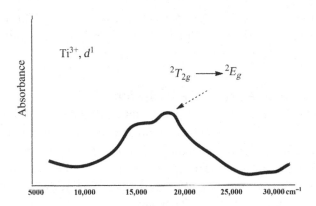

Figure 2.27 Spectrum of an aqueous solution of Ti^{3+}, a d^1 system.

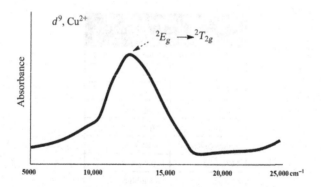

Figure 2.28 The electronic spectrum of Cu^{2+}, a d^9 system.

2.8.2 Electronic Spectrum of a d^9 System

The electronic spectrum of Cu^{2+} is shown in Fig. 2.28. According to the *hole formalism*, it should be the reverse of the d^1 system. Thus the transition will be $^2E_g \rightarrow {}^2T_{2g}$.

2.8.3 Electronic Spectrum of a d^2 System

The electronic spectrum of V^{3+}, a d^2, system is shown in Fig. 2.29. The Tanabe-Sugano diagram for this system is shown in Fig. 2.25, where the ground state is $^3T_{1g}(F)$ and the excited states with the same multiplicity are $^3T_{2g}(F)$, $^3T_{1g}(P)$, and $^3A_{2g}(F)$. Thus the spin-allowed transitions will be $^3T_{1g}(F) \rightarrow {}^3T_{2g}(F)$, $^3T_{1g}(F) \rightarrow {}^3T_{1g}(P)$, and $^3T_{1g}(F) \rightarrow {}^3A_{2g}(F)$. Fig. 2.29 shows two transitions and these are the first two of the three allowed transitions. The third one, namely $^3T_{1g}(F) \rightarrow {}^3A_{2g}(F)$ is not visible in this region because of its high energy. It might have been shifted to the UV region.

2.8.4 Electronic Spectrum of a d^8 System

The electronic spectrum of an aqueous solution of the d^8 Ni^{2+} ion is given in Fig. 2.30. According to the *hole formalism*, it is an inverted d^2 system. When we refer to the Tanabe-Sugano diagram for the d^8 system, we see that $^3A_{2g}(F)$ is the ground state. The excited states with the same multiplicity are $^3T_{2g}(F)$, $^3T_{1g}(F)$, and $^3T_{1g}(P)$. Thus three spin-allowed (multiplicity allowed) transitions are possible. They are

$$^3A_{2g}(F) \rightarrow {}^3T_{2g}(F) \quad {}^3A_{2g}(F) \rightarrow {}^3T_{1g}(F) \quad {}^3A_{2g}(F) \rightarrow {}^3T_{1g}(P)$$

Fig. 2.30 shows all the three transitions.

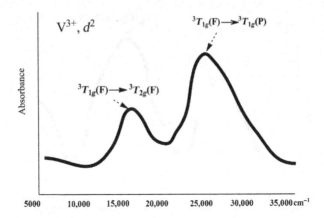

Figure 2.29 The electronic spectrum of V^{3+}, a d^2 system.

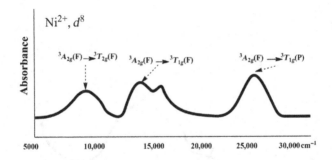

Figure 2.30 Electronic spectrum of Ni^{2+} ion in aqueous solution.

2.8.5 Electronic Spectrum of a d^3 System

The electronic spectrum of Cr^{3+} is shown in Fig. 2.31. Cr^{3+} is a d^3 system. According to the *hole formalism*, it is same as an *inverted d^2* system. Thus the allowed transitions will be $^4A_{2g} \rightarrow {}^4T_{2g}(F)$, $^4A_{2g} \rightarrow {}^4T_{1g}(F)$, and $^4A_{2g} \rightarrow {}^4T_{1g}(P)$.

2.8.6 Electronic Spectrum of a d^7 System

An aqueous solution of Co^{2+} is pink in color and its electronic spectrum is shown in Fig. 2.32. The Tanabe-Sugano diagram for a d^7 system is given in Fig. 2.33 and the transitions are assigned as

$$^4T_{1g}(F) \rightarrow {}^4T_{2g}(F), \quad {}^4T_{1g}(F) \rightarrow {}^4A_{2g}(F), \quad \text{and} \quad {}^4T_{1g}(F) \rightarrow {}^4T_{1g}(P)$$

as shown in Fig. 2.32. In the weak-field case, it is the same as the d^2 system (Fig. 2.23B) (*hole formalism*) and only the multiplicity will change.

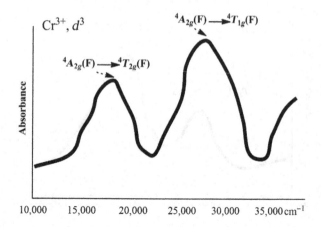

Figure 2.31 The electronic spectrum of an aqueous solution of Cr^{3+}.

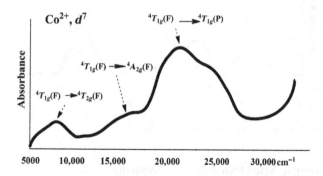

Figure 2.32 The electronic spectrum of Co^{2+} aqueous solution.

2.8.7 Electronic Spectrum of a d^4 System

The electronic spectrum of Cr^{2+}, which is a d^4 system, is given in Fig. 2.34. Since it is an aqueous solution, the ligand is H_2O and it is a weak ligand. Applying the *hole formalism*, it can be considered as d^{5-1} and hence an inverted d^1 system. Thus, it is the reverse of Fig. 2.23A. Table 2.12 shows that the free-ion term is ^5D and the ground term is ^5E_g. Hence, the transition should be $^5E_g \rightarrow {}^5T_{2g}$. This is confirmed by the Tanabe-Sugano diagram for a d^4 system (Fig. 2.26). Looking at the left-hand side of the vertical line in the diagram, we find that the spin-allowed transition is $^5E_g \rightarrow {}^5T_{2g}$.

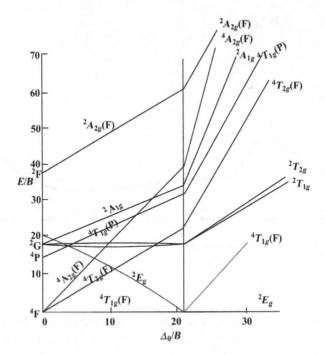

Figure 2.33 The Tanabe-Sugano diagram for a d^7 system.

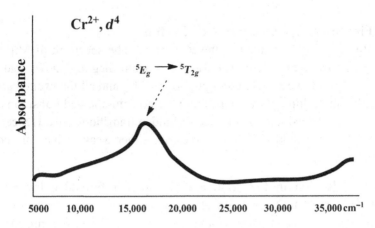

Figure 2.34 Electronic spectrum of Cr^{2+} aqueous solution.

2.8.8 Electronic Spectrum of a d^6 System

The spectrum of Fe^{2+}, which is a d^6 system, is shown in Fig. 2.35. This is a d^{5+1} system. That is, according to the *hole formalism*, the splitting and

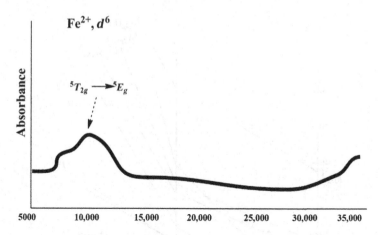

Figure 2.35 Electronic spectrum of an aqueous solution of Fe^{2+}.

energy levels are the same as those of the d^1 system but the multiplicity will change. Then the transition will be $^5T_{2g} \rightarrow {}^5E_g$. This is also supported by the Tanabe-Sugano diagram for the d^1 system.

2.8.9 Electronic Spectrum of a d^5 System

Both Mn^{2+} and Fe^{3+} belong to the d^5 system. The spectrum of Mn^{2+} in water is shown in Fig. 2.36. The Tanabe-Sugano diagram shows that the ground state for a weak-field configuration is $^6A_{1g}$ and all the excited states have different multiplicities. Thus, there are no spin-allowed transitions for the d^5 system. Transitions, if any, are forbidden transitions only. Hence, the color of the Mn^{2+} and Fe^{3+} complexes are either weak or have no color other than white.

The bands seen in Fig. 2.36 are due to spin-forbidden transitions. Unusually, some bands are sharp or narrower because the energies of the corresponding excited states are not varying with Δ_o. These will run parallel to the x-axis. Another factor for the narrow band may be that the spin-orbit coupling is not splitting the excited states.

The same argument holds good for $[Fe(H_2O)_6]^{3+}$. However, the solution is yellow. This color is not due to the ligand field transitions but due to the hydrolysis and CT bands of polymeric species formed during hydrolysis.

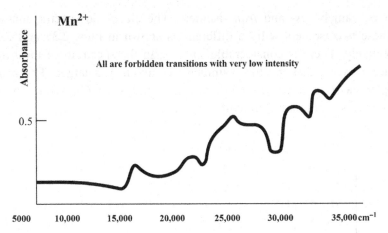

Figure 2.36 The electronic spectrum of an aqueous solution of Mn^{2+}.

2.9 *CIS-* AND *TRANS-*COMPLEXES

[Co(en)$_3$]$^{3+}$ is a complex with cubic symmetry. It is a d^6 strong-field system and from the corresponding Tanabe-Sugano diagram (right side of the vertical line), two transitions are expected and they would be assigned as $^1A_{1g} \rightarrow {}^1T_{1g}$ and $^1A_{1g} \rightarrow {}^1T_{2g}$. This is shown in Fig. 2.37.

Now, let us replace one (en) with two F$^-$. This substitution of ligand leads to reduction of symmetry from cubic to D_{4h}. Also, we will have two

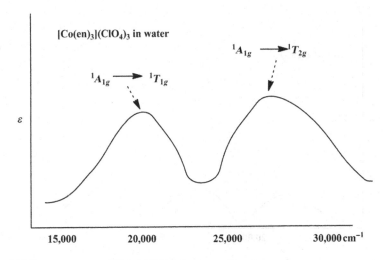

Figure 2.37 Electronic spectrum of [Co(en)$_3$](ClO$_4$)$_3$ in water.

isomers, namely *cis-* and *trans-*isomers. The electronic spectra obtained for these two isomers will be different as shown in Figs. 2.38 and 2.39, respectively. There is considerable splitting in the spectrum of the *trans-*isomer showing that the low symmetry component is larger. The *trans-*complex has D_{4h} symmetry, while the *cis-*complex has C_{2v} symmetry and hence their spectra look different.

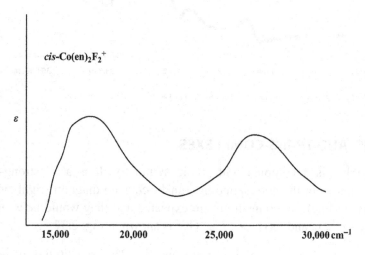

Figure 2.38 Electronic spectrum of cis-[Co(en)₂F₂]⁺.

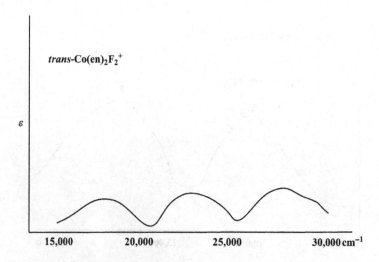

Figure 2.39 Electronic spectrum of trans-[Co(en)₂F₂]⁺.

2.10 RULE OF AVERAGE ENVIRONMENT

A complex can maintain some amount of cubic symmetry even when there is reduction in symmetry. Consider an octahedral complex having mixed ligands. Thus its symmetry will be lowered from cubic symmetry. However, the rule of average environment says that when we have an octahedral complex with the mixture of ligands, the amount of cubic field component present in the complex will be a weighted average of the cubic ligand fields in the pure complexes without any mixture of ligands. Thus, the amount of cubic field present in the complex $[Co(en)_2F_2]^+$ will be derived from the complexes $[Co(en)_3]^{3+}$ and $[CoF_6]^{3-}$. Δ_o for $[Co(en)_3]^{3+}$ and $[CoF_6]^{3-}$ are 23,000 and 13,000 cm^{-1}, respectively. Then, according to the rule of average environment,

$$\Delta_o(MA_nB_{6-n}) = \frac{[n\Delta_0(MA_6) + (6-n)\Delta_o(MB_6)]}{6} \ cm^{-1} \qquad (2.23)$$

Thus in the case of $[Co(en)_2F_2]^+$,

$$\Delta_o = \frac{(4 \times 23{,}000 + 2 \times 13{,}000)}{6} = 20{,}000 \ cm^{-1} \qquad (2.24)$$

This agrees reasonably well with the experimental value of 21,000 cm^{-1}. Thus, the rule of average environment helps us to calculate the Δ_o and other ligand field parameters for mixed ligand complexes with confidence.

2.11 NEPHELAUXETIC EFFECT

Nephelauxetic means 'cloud expanding'. Here, 'cloud' refers to the orbital. Thus, the nephelauxetic effect means 'expansion of the orbital' formed by the overlap of metal and ligand orbitals during complex formation.

When the metal d-orbitals overlap with the ligand orbital, a bigger MO is formed. Since electrons are occupying a bigger orbital now, the distance between them increases and hence the repulsion decreases. This is called the nephelauxetic effect.

In other words, the interelectronic repulsion is always less in the complex compared to the free ion. The interelectronic repulsion is denoted as B' in the complex and B in the free ion. Then, the ratio between these two is called the nephelauxetic ratio (β):

$$\beta = \frac{B'}{B}$$

β is always less than 1.

B' is determined from the spectrum of the complex and the appropriate Tanabe-Sugano diagram as described earlier. B for metal ions is readily available. Hence, β can be determined easily for complexes.

This nephelauxetic ratio is an important parameter because it gives the degree of covalence in the metal-ligand bond.

2.12 SPECTRA OF TETRAHEDRAL COMPLEXES

While octahedral complexes are large in number, tetrahedral complexes are relatively scarce. The interpretation of the spectra is the same in both cases. Certain points are to be noted in tetrahedral complexes:

1. The spectral bands are more intense than those of the octahedral complexes due to d-p mixing.
2. Splitting is less compared to octahedral complexes according to:

$$\Delta_t = \frac{4}{9}\Delta_o \tag{2.25}$$

3. Effects of vibronic interaction, spin-orbit coupling, and low-symmetry ligand fields are stronger in tetrahedral complexes compared to octahedral complexes.
4. Spectral interpretation is based on Tanabe-Sugano diagrams. A d^n tetrahedral system is interpreted with a diagram for a d^{10-n} system.

2.13 CT SPECTRA (CHARGE TRANSFER)

CT refers to the transfer of electrons from ligand to metal or vice versa. Hence, these are also known as redox spectra. If the CT bands appear in the visible region, the complex will be intensely colored.

If the metal is a very good oxidizing agent and the ligand group is a good reducing agent, then electrons (charge) can easily flow from ligand to metal and the CT spectrum will have low energy. These CT bands are assigned as transitions between MOs of the complex. If the metal and ligand orbital wave functions mix extensively, the CT transitions and d-d transitions (ligand field transitions) cannot be distinguished.

2.13.1 How to Distinguish Between CT and d-d Transitions

The energy of a CT transition depends on the solvent polarity, while that of a d-d transition does not. Hence, when the solvent polarity is changed, the position of a CT band will change, while that of a d-d transition band will not change. Thus CT and d-d transitions can be distinguished.

2.13.2 Types of CT Transitions

There are two types of CT transitions, namely LMCT (Ligand to Metal Charge Transfer) and MLCT (Metal to Ligand Charge Transfer) transitions.

LMCT Transition

Here, charge (electrons) flows from ligand orbitals to metal π-orbitals. This type of CT transition will occur when the ligands have lone pairs of electrons or when the metals have low-lying π-orbitals. For example, CdS is yellow in color and it is due to transfer of electrons from S^{2-} (π-orbital) to Cd^{2+} ($5s$ orbital). HgS is red in color and this color is due to the CT from S^{2-} (π-orbital) to Hg^{2+} ($6s$ orbital). In the same way, ochres, which are iron oxides, are colored red and yellow due to the transfer of charge from O^{2-} to Fe ($3d$ orbitals).

This type of transition usually occurs when the metals are in high oxidation states because they can be easily reduced to lower oxidation states as in the case of MnO_4^-. Here, the lone pair on oxygen is transferred to the low-lying metal orbital.

MLCT Transition

These are usually observed in complexes containing aromatic ligands because they will have low-lying π^*-orbitals. In these complexes, metals will be in low oxidation states and hence their d-orbitals will have high energy. Under these circumstances, the CT transition will occur at low energy. Complexes containing ligands such as diimines, 2,2'-bipyridine, and 1,10-phenanthroline exhibit MLCT transitions.

EXERCISES

Exercise 2.1. Identify the symmetry elements in $PtCl_4^{2-}$

Exercise 2.2. What is a microstate?

Exercise 2.3. What is the degeneracy of the term ^3F? (*Hint:* $(2S+1)(2L+1)$)

Exercise 2.4. What is the degeneracy of the state $^3F_2$? (*Hint:* $2J+1$)

Exercise 2.5. How many microstates are possible for a d^2 configuration? ($Hint: N = \dfrac{N_l!}{[x!(N_l - x)!]}$)

Exercise 2.6. Derive the symmetry of the complex $Co(en)_3^{3+}$.

Exercise 2.7. Derive the microstates and construct a microchart for the excited state of Be.

Exercise 2.8. Suppose a term splits into ^1S and ^3P, which will have lower energy? Why?

Exercise 2.9. Out of the two terms, ^3P and ^3F, which one will have lower energy and why?

Exercise 2.10. A configuration gives rise to two terms, ^3F and ^3P. What is that configuration? Of the two terms, which one will have lower energy and why?

Exercise 2.11. What is the ground state term for a d^4 configuration?

Exercise 2.12. In an octahedral environment, identify the ground terms for (a) weak-field and (b) strong-field d^6 configurations.

Exercise 2.13. What is the ground state for the complex $[V(H_2O)_6]^{3+}$? What are the excited states? Identify the allowed transitions.

Exercise 2.14. For a complex, $[V(H_2O)_6]^{n+}$, transitions have been assigned as $^3T_{1g}(F) \rightarrow ^3T_{2g}$, $^3T_{1g}(F) \rightarrow ^3T_{1g}(P)$, and $^3T_{1g}(F) \rightarrow ^3A_{2g}$ around 18,000, 25,000, and 37,000 cm^{-1}, respectively. What is the oxidation state of V? Is there any possibility of CT transition in this complex? If so, is it LMCT or MLCT? If there is a CT transition and if this also happens around 37,000 cm^{-1}, how can we say whether it is a CT or d-d transition? Try to find Δ_o and the nephelauxetic ratio for this complex.

Exercise 2.15. Let us consider a complex of d^4 high-spin as well as low-spin complex. Will either or both undergo Jahn-Teller distortion? If so, in which case will the distortion be greater and why?

Exercise 2.16. Consider an octahedral complex. Draw the energy levels with the symmetries of the orbitals involved. Let it undergo z-out distortion giving a square planar environment. Show how the energy levels are split together with the symmetries of orbitals. How do you find the symmetries of orbitals in different environments?

Exercise 2.17. In a cobalt(II) octahedral complex, two transitions are observed. What are they? Of these two, one is weak. Identify that transition and give the reason for its weakness. (*Hint*: Number of electrons excited)

Exercise 2.18. Consider two octahedral complexes. In both cases, CT transition takes place. In one case, the metal is in the +7 state and in the other metal it is in the +1 state. In which case will MLCT and in which case LMCT transitions occur? Give reasons.

Exercise 2.19. Consider the two octahedral complexes, $Co(NH_3)_6^{3+}$ and $Co(CN)_6^{3-}$. In which case is MLCT transition possible? Why?

Exercise 2.20. All the following complexes show CT transitions: VO_4^{3-}, CrO_4^{2-}, and MnO_4^-. λ increases from V to Mn complex. That is, MnO_4^- has the longest wavelength. Account for this observation.

REFERENCES

[1] J.E. Huheey, E.A. Keitler, R.L. Keitler, O.K. Medhi, Inorganic Chemistry, Pearson, Delhi, 2006.

[2] G. Wulfsberg, Inorganic Chemistry, first ed., Viva Books, Chennai, 2005.

[3] G.L. Miessler, D.A. Tarr, Inorganic Chemistry, third ed., Pearson, New Delhi, 2012.

[4] F.A. Cotton, Chemical Applications of Group Theory, second ed., Wiley Eastern, New Delhi, 1986.

[5] A.B.P. Lever, Electronic Spectroscopy, Elsevier, New York, 1984.

[6] A.K. Brisdon, Inorganic Spectroscopic Methods, Oxford Science, New York, 2010.

CHAPTER 3

IR Spectroscopy

3.1 SOME FUNDAMENTALS

Already we have seen in Chapter 1 that the number of vibrations for a linear molecule will be $3N - 5$ and that for nonlinear molecules will be $3N - 6$, where N is the total number of atoms in the molecule. $3N$ is the total number of degrees of freedom, that is, the number of translational motion + number of rotations + number of vibrations. For all molecules, linear or nonlinear, the number of translations will be equal to 3 because the translational motion is along the three axes, X-, Y-, and Z-axes. However, the difference comes only in the rotation: the number of rotations = 2 for linear molecules and 3 for nonlinear molecules. Thus, the difference in the number of vibrations of linear and nonlinear molecules is due to the difference in the number of rotations. These vibrations are called *fundamental vibrations*.

3.1.1 Fundamental Vibrations or Normal Vibrations

Those vibrations in which all the atoms move in phase and also move with the same frequency are called *normal vibrations* [1].

Nonlinear Molecules

H_2O is V-shaped, that is, nonlinear. Hence, the number of fundamental vibrations is equal to $3N - 6 = 3 \times 3 - 6 = 3$. These are the symmetric stretching, antisymmetric stretching, and symmetric bending vibrations as shown in Fig. 3.1A-C.

Linear Molecules

The number of fundamental vibrations for a linear molecule such as CO_2 will be equal to $3N - 5 = 3 \times 3 - 5 = 4$ vibrations. These are shown in Fig. 3.2A-C.

3.1.2 Overtone and Combination Bands

Overtones

Real molecules do not exhibit harmonic motion but exhibit anharmonic motion. Because of this, in addition to fundamental or normal vibrations,

Spectral Methods in Transition Metal Complexes. http://dx.doi.org/10.1016/B978-0-12-809591-1.00003-7

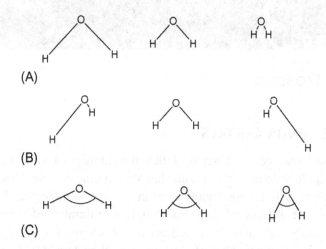

Figure 3.1 *Fundamental vibrations of the water molecule. (A) Symmetric stretching vibrations. (B) Asymmetric stretching vibrations. (C) Symmetric bending vibrations.*

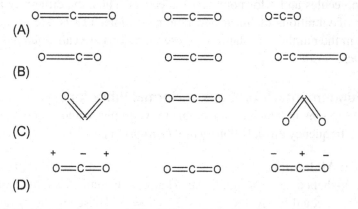

Figure 3.2 *Normal modes of vibration of CO_2. (A) Symmetric stretching vibrations. (B) Asymmetric stretching vibrations. (C) In-plane bending. (D) Out-of-plane bending. "+", the atom is coming above the plane. "−", the atom goes below the plane.*

overtones appear. If ν_1 is the fundamental frequency, the first overtone appears at $2\nu_1$, the second overtone appears at $3\nu_1$ and so on. For example, if the fundamental vibration appears at $800 \ cm^{-1}$, its first overtone appears at $2 \times 800 = 1600 \ cm^{-1}$, the second overtone appears at $3 \times 800 = 2400 \ cm^{-1}$ and so on. Thus, these overtones may complicate the spectrum by their appearance as additional bands. However, their intensities decrease rapidly.

Combination Bands

Combination bands result due to the addition of two or more fundamental frequencies or overtones but intensities will be very small. These can be represented as $v_1 + v_2, v_1 + v_2 + v_3, 2v_1 + v_2$, etc.

Difference Bands

Difference bands are obtained by subtracting one frequency from the other. These can be represented as $v_1 - v_2, v_1 + v_2 - v_3, 2v_1 - v_2$, etc.

3.2 SELECTION RULE FOR IR SPECTRA

If a vibration is to be seen in the IR spectrum (ie, IR-active vibration), there should be a change in dipole moment. This is the selection rule for IR spectra. In other words, the molecule should have a dipole moment initially. Hence, homodiatomic molecules such as H_2, O_2, Cl_2, N_2, etc., which have no dipole moment will be IR inactive, while heteronuclear diatomic molecules such as HCl will be IR active because they have dipole moments. In the case of polyatomic molecules, the IR-active vibrations can be predicted from the appropriate character table.

Example: H_2O has C_{2v} symmetry. Hence, the C_{2v} character table is consulted. It is given in Table 3.1. The dipole moments can be compared to the translations along the three axes. Therefore, the IR-active vibrations for a water molecule or any molecule having C_{2v} symmetry have the symmetries A_1, B_1, and B_2 noted against z, x, and y, respectively. Thus, it should be noted that x, y, and z appearing in the character tables represent *translational* motions along the three axes, dipole moments along the three axes and also the symmetries of the p_x, p_y, and p_z orbitals.

Table 3.1 C_{2v} Character Table						
C_{2v}	E	C_2	$\sigma_v(xz)$	$\sigma'_v(yz)$		
A_1	1	1	1	1	z	x^2, y^2, z^2
A_2	1	1	-1	-1	R_z	xy
B_1	1	-1	1	-1	x, R_y	xz
B_2	1	-1	-1	1	y, R_x	yz

3.3 SELECTION RULE FOR RAMAN SPECTRA

The selection rule for Raman spectra is that there should be a change in *polarizability* of the molecule. Polarizability is the distortion in the electron cloud. Homonuclear diatomic molecules will be Raman active.

3.4 RULE OF MUTUAL EXCLUSION

According to the *mutual exclusion principle,* if a molecule has center of symmetry, then IR-active vibrations will be Raman inactive and Raman-active vibrations will be IR inactive. The IR- and Raman-active vibrations can be easily identified from the corresponding character table. The IR-active vibrations will have "u" symmetry while Raman-active vibrations will have "g" symmetry. As an example, in the C_{2h} character table shown in Table 3.5, we have A_g, B_g, A_u, and B_u terms. Of these, vibrations with A_g and B_g symmetries will be Raman active, while those with A_u and B_u symmetries will be IR active.

3.5 IR SPECTRA AND INORGANIC COMPOUNDS

We can predict the number of IR-active and Raman-active vibrations in a molecule with the help of SALC and group theory as explained below.

3.5.1 SALC and Prediction of IR- and Raman-Active Bands

Group theory alone is not sufficient to find the normal modes of vibration in a molecule. If an irreducible representation has only one vibration, group theory is sufficient to identify the normal modes of vibration and is called *symmetry determined.* If the irreducible representation has more than one vibration, group theory gives symmetry adapted vectors (basis) and these vectors are mixed to form the normal modes. The same applies to molecular orbitals (MOs) as well. However, the basis set is atomic orbitals (AOs) instead of vectors in this case.

3.5.2 What Is SALC and Why Is It Necessary?

SALC is Symmetry Adapted Linear Combination. Molecular geometry and character tables help to find the number of vibrational modes for each symmetry type (ie, for each irreducible representation). If an irreducible representation has only one vibration, then the normal mode represented by that is called "symmetry determined." However, if an irreducible representation has more than one vibration, then the vectors representating these are mixed to form *normal modes*. Hence, we must know how to form "Symmetry Adapted Linear Combinations" of the basis functions such as AOs for constructing MOs or basis vectors in the case of vibrations.

Methods to Obtain SALCs

There are two methods to obtain SALCs:

1. basis vector method and
2. projection operator method.

Basis Vector Method

Derivation of SALCs based on the basis vector method involves the following steps:

1. The point group of the molecule is derived first.
2. Then the corresponding character table of the point group is consulted.
3. The reducible representation is derived.
4. This is reduced to its components using the reduction formula.
5. After that normalization is done.
6. Finally SALCs are obtained.

As an example, let us take the water molecule. Its point group is C_{2v}. The C_{2v} character table is given in Table 3.1. The *reducible representation* of the water molecule is derived by employing the symmetry operations as shown in Figs. 3.3–3.5.

Thus, the reducible representation of the water molecule obtained is given in Table 3.2.

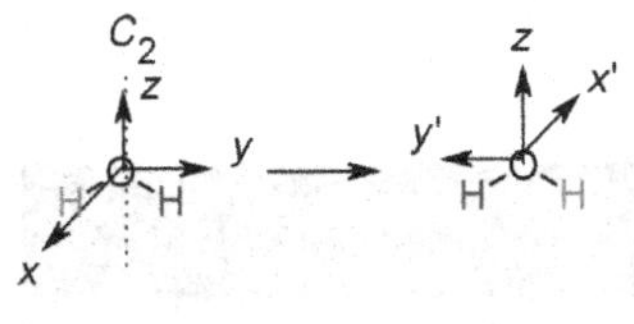

Figure 3.3 C_2 operation on a water molecule.

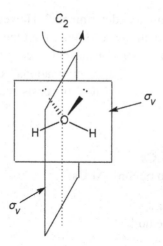

Figure 3.4 σ$_v$ operation on the water molecule.

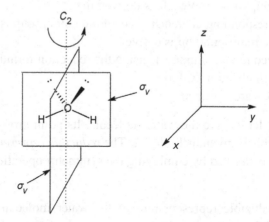

Figure 3.5 Notation of axes and planes.

Table 3.2 Reducible Representation of Water				
C_{2v}	E	C_2	$\sigma_v(xz)$	$\sigma_v(yz)$
Γ	9	−1	1	3

The reduction formula is given as:

$$a_i = \frac{1}{h}\Sigma n_i \chi_i \chi_r \qquad (3.1)$$

where a_i is the number of times a symmetry species occurs in the reducible representation, "h" is the order of the reducible representation and is the total number of symmetry operations in that group. This is obtained by adding the coefficients of the symmetry operations in the character table; "n_i" is the number of symmetry operations in that type, χ_i is the character of the irreducible representation, and χ_r is the character of the reducible representation. The reduction is carried out as shown in Eq. (3.1).

$$A_1 = \frac{1}{4}[(1)(1)(9) + (1)(1)(-1) + (1)(1)(1) + (1)(1)(3)]$$

$$= \frac{1}{4}[12] = 3$$

$$A_2 = \frac{1}{4}[(1)(1)(9) + (1)(1)(-1) + (1)(-1)(1) + (1)(-1)(3)]$$

$$= \frac{1}{4}[4] = 1$$

$$B_1 = \frac{1}{4}[(1)(1)(9) + (1)(-1)(-1) + (1)(1)(1) + (1)(-1)(3)]$$

$$= \frac{1}{4}[8] = 2$$

$$B_2 = \frac{1}{4}[(1)(1)(9) + (1)(-1)(-1) + (1)(-1)(1) + (1)(1)(3)]$$

$$= \frac{1}{4}[12] = 3$$

Thus, we get the following result finally after reduction:
Translation + rotation + vibration = $3A_1 + A_2 + 2B_1 + 3B_2$
Translation = $A_1 + B_1 + B_2$
Rotation = $A_2 + B_1 + B_2$
Vibration = $2A_1 + B_2$ (translation and rotation are subtracted from the total)

Constructing SALCs
The symmetries of the stretching modes found above are: A_1 symmetric stretching and B_2 antisymmetric stretching. These are shown in Fig. 3.6.

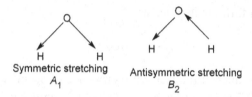

Figure 3.6 Symmetric and antisymmetric stretchings.

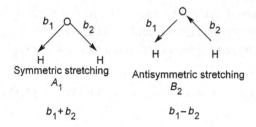

Figure 3.7 Basis set for SALC.

3.5.3 Choosing the Basis Set and Linear Combination

Now, the basis set is chosen. In this case, we have vectors (arrows) along the bonds as shown in Fig. 3.7.

3.5.4 Expression for SALCs

The SALCs are the sum over all the basis functions as follows:

$$\phi_i = \sum_j c_{ij} b_{ij} \tag{3.2}$$

where ϕ_i is the ith SALC function, b_j is the jth basis function, and c_{ij} is a coefficient which tells the contribution of b_j to ϕ_i.

3.5.5 Normalization

In order to normalize, the above expression is to be multiplied by the normalization constant, N. It is the inverse of the square root of the sum of the squares of the coefficients within the expression. The expression for normalization is

$$N = \frac{1}{\sqrt{\sum_{j=1}^{n} c_{ij}^2}} \tag{3.3}$$

This normalization of SALC ensures that the magnitude of the SALC is unity. Hence, the dot product of any SALC with itself will be equal to one. The normalized SALCs are

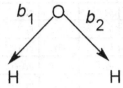

Figure 3.8 Basis vectors.

$$\frac{1}{\sqrt{2}}(b_1 + b_2)$$

$$\frac{1}{\sqrt{2}}(b_1 - b_2)$$

Projection Operator Method
This method involves the following steps:

Step 1: Basis vectors are shown in Fig. 3.8.

The basis vector is represented as v and the projection vector is represented as T_j for the jth symmetry operation.

3.5.6 Transformations of the Basis Vectors
The transformations of the basis vectors are shown below:

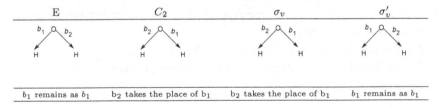

E	C_2	σ_v	σ'_v
b_1 remains as b_1	b_2 takes the place of b_1	b_2 takes the place of b_1	b_1 remains as b_1

3.5.7 SALC Functions and Basis Sets
The SALC functions are the collective transformations of the basis sets represented by ϕ_i and are obtained from Eq. (3.4). $\chi_i(j)$ is the character of the ith reducible representation and the jth symmetry operation:

$$\phi_i = \sum_j \chi_i(j) T_j v \tag{3.4}$$

	E	C_2	σ_v	σ_v'	SUM
$T_j b_1$	b_1	b_2	b_2	b_1	
$A_1 T_j b_1$	$b_1(1)$	$b_2(1)$	$b_2(1)$	$b_1(1)$	$2b_1 + 2b_2$
$A_2 T_j b_1$	$b_1(1)$	$b_2(1)$	$b_2(-1)$	$b_1(-1)$	0
$B_1 T_j b_1$	$b_1(1)$	$b_2(-1)$	$b_2(1)$	$b_1(-1)$	0
$B_2 T_j b_1$	$b_1(1)$	$b_2(-1)$	$b_2(-1)$	$b_1(1)$	$2b_1 - 2b_2$

3.5.8 Normalization

$$\phi(A_1) = \frac{1}{\sqrt{2}}(b_1 + b_2)$$

$$\phi(B_2) = \frac{1}{\sqrt{2}}(b_1 - b_2)$$

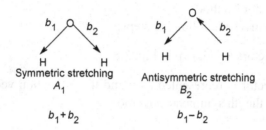

Symmetric stretching
A_1

$b_1 + b_2$

Antisymmetric stretching
B_2

$b_1 - b_2$

3.5.9 Interpretation of SALCs
- Both the methods give the same results.
- Irreducible representations corresponding to the stretching modes alone will produce nonzero SALC.
- Nonvibrational modes produced zero SALCs.
- All irreducible representations of the points groups were used.

3.5.10 Applications
- From the SALCs of a molecule, all stretching modes can be identified.
- This provides clear understanding of the vibrational spectra.
- SALCs help in determining the geometry of a molecule.
- SALCs can distinguish between different isomers of a molecule, for example, *ortho-* and *para-*difluorobenzenes.

Examples
Example 1: *trans*-N_2F_2. Its structure is shown in Fig. 3.9.

Figure 3.9 Structure of trans-N_2F_2.

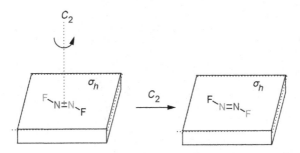

Figure 3.10 Effect of C_2 operation on trans-N_2F_2.

Derivation of the Point Group of *trans*-N_2F_2

1. This has a C_2 axis perpendicular to the molecular plane, ie, the plane of the paper. That is, rotation by $180°$ gives an indistinguishable structure as shown in Fig. 3.10.
2. It has a σ_h plane as shown in Fig. 3.11.
3. Hence, the point group is C_{2h}.

C_{2h} Character Table

The C_{2h} character table is shown in Table 3.3.

Derivation of the Reducible Representation

The different steps involved are listed below:

1. Base vectors are assigned appropriately for each atom: arrows on the three axes for translation or vibration, curved arrows on the axes for rotation, etc., as the case may be.
2. For each operation, it is found how many atoms in the molecule remain *undisturbed*. Only the vectors on these atoms are considered for determining the characters.
3. Firstly, the vectors on one undisturbed atom are considered and the character for a particular symmetry operation is derived. Then, it is multiplied by the total number of undisturbed atoms to find the total character for that particular symmetry operation.

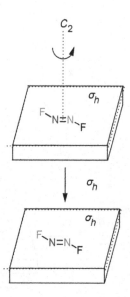

Figure 3.11 Effect of σ_h operation on trans-N_2F_2.

Table 3.3 C_{2h} Character Table						
C_{2h}	E	C_2	i	σ_h		
A_g	1	1	1	1	R_z	x^2, y^2, z^2, xy
B_g	1	−1	1	−1	R_x, R_y	xz, yz
A_u	1	1	−1	−1	z	
B_u	1	−1	−1	1	x, y	

4. This is repeated for other symmetry operations and, finally, the reducible representation is obtained.

Identity Operation, **E**   There are four atoms in the molecule and three vectors for each atom. Hence, in total, there will be *12* vectors. Since all the four atoms remain undisturbed, the vectors associated with them also remain undisturbed. Hence, $\chi(\mathbf{E}) = 12$.

C$_2$ *Operation*   The effect of **C$_2$** operation is shown in Fig. 3.10. Since all the four atoms are displaced from their original positions, $\chi(\mathbf{C}_2) = 0$.

i Operation   In this inversion operation also, all the four atoms are displaced from their original positions. Hence, $\chi(i) = 0$.

σ_h *Operation*   The effect of σ_h operation is shown in Fig. 3.11. All the four atoms remain undisturbed. Hence, all the *12* vectors (arrows on the axes) remain undisturbed and therefore $\chi(\sigma_h) = 12$.

Reducible Representation   The reducible representation derived is given in Table 3.4.

Obtaining the Irreducible Representation   We will be reducing the reducible representation using the reduction formula. For convenience, let us write the reducible representation as the last row in the character table as shown in Table 3.5. The reducible representation is obtained as shown below:

$$A_g = \frac{1}{4}[(12)(1)(1) + (0)(1)(1) + (0)(1)(1) + (4)(1)(1)]$$

$$= \frac{1}{4}(12 + 4) = 4$$

$$B_g = \frac{1}{4}[(12)(1)(1) + (0)(1)(-1) + (0)(1)(1) + (4)(1)(-1)]$$

$$= \frac{1}{4}(12 - 4) = 2$$

$$A_u = \frac{1}{4}[(12)(1)(1) + (0)(1)(1) + (0)(1)(-1) + (4)(1)(-1)] \quad (3.5)$$

$$= \frac{1}{4}(12 - 4) = 2$$

$$B_u = \frac{1}{4}[(12)(1)(1) + (0)(1)(-1) + (0)(1)(-1) + (4)(1)(1)]$$

$$= \frac{1}{4}(12 + 4) = 4$$

Thus, $\Gamma = 4A_g + 2B_g + 2A_u + 4B_u$

Table 3.4 Reducible Representation for *trans*-N_2F_2

C_{2h}	E	C_2	i	σ_h
Γ	12	0	0	4

Table 3.5 C_{2h} Character Table With the Reducible Representation for *trans*-N_2F_2						
C_{2h}	E	C_2	i	σ_h		
A_g	1	1	1	1	R_z	x^2, y^2, z^2, xy
B_g	1	−1	1	−1	R_x, R_y	xz, yz
A_u	1	1	−1	−1	z	
B_u	1	−1	−1	1	x, y	
Γ	12	0	0	4		

Interpretation of the Result The reducible representation contains the A_g symmetry *four* times, B_g symmetry *two* times, A_u symmetry *two* times, and B_u symmetry *four* times. From the character table it can be seen that A_g represents rotation along the z-axis, R_z, B_g symmetry represents that along the x- and y-axes, R_x and R_y. A_u symmetry represents translation along the z-axis and B_u represents translation along the x- and y-axes.

Number of IR- and Raman-Active Vibrations The total number of vibrations for a nonlinear molecule = $(3N - 6)$ and for a linear molecule it is $(3N - 5)$. $3N$ represents the number of translations + number of rotations + number of vibrations. The numbers 6 and 5 represent the number of rotations + number of translations. The molecule under consideration, namely *trans*-N_2F_2 is a nonlinear molecule. Hence, the number of vibrations = $3 \times 4 - 6 = 6$. With this background, now let us use the derived reducible and irreducible representations. The total number of vibrations (IR + Raman) is given by Eq. (3.6).

$$\text{Translation + rotation + vibration} = 4A_g + 2B_g + 2A_u + 4B_u$$
$$\text{Translation} = A_u + 2B_u$$
$$\text{Rotation} = A_g + 2B_g$$
$$\therefore \text{Vibration} = [(4A_g + 2B_g + 2A_u + 4B_u) \tag{3.6}$$
$$- (A_u + 2B_u) - (A_g + 2B_g)]$$
$$\text{Vibration} = 3A_g + A_u + 2B_u$$

Thus, there are *six* vibrations in total. This agrees with the $(3N - 6)$ formula. Thus the number of vibrations in a molecule has been found using the reducible and irreducible representations.

IR-Active Vibrations   The IR-active vibrations can be found easily from the character table. Those symmetries, which have subscript 'u' are IR active. Thus the three vibrations with $A_u + 2B_u$ symmetries will be IR active.

Raman-Active Vibrations   The Raman-active vibrations can be found from the character table. Those symmetries with subscript 'g' will be Raman active. Thus the three vibrations with $A_g + 2B_g$ symmetries will be Raman active.

Another feature of IR is that when we can expect more than one structure for a given complex; the correct structure can be found using the symmetry, character table, and SALC discussed above.

Example: In order to understand how IR is useful in determining the structure of complexes, we can consider $NiCl_4$. This can have either *tetrahedral* (T_d) or *square planar* (D_{4h}) structures. The number of IR-active vibrations are derived using T_d and D_{4h} character tables and SALC as explained above. The results are compared with that obtained from the actual IR spectrum recorded for the complex. The structure for which both the predicted and actual data agree is the correct structure of the complex. In this way, it can be shown that $NiCl_4$ is **tetrahedral** and not square planar.

3.5.11 Interpretation of IR Spectra

Thus, the IR spectrum of a molecule will contain a large number of bands and will be complicated. Hence, it is not possible or necessary to interpret all the bands in the IR spectrum. Instead, the bands corresponding to the different functional groups (Table 3.6) are considered. In order to find whether a complex is formed or not, the following points are considered:

1. Is there any shift in the position of the bands of the functional group of the ligand?
2. Are additional bands seen in the complex compared to that of the free ligand?
3. Is there any change in the intensity of the bands of the functional groups in the complexes compared to the ligands?

All these points are explained in the following sections.

Table 3.6 IR Absorptions of Important Functional Groups

Frequency (cm^{-1})	Bond	Str/Bend	Functional Group
3640–3610 (s, sh)	O–H (free)	Str	Alcohols/phenols
3500–3200 (s, b)	O–H (H-bonded)	Str	Alcohols/phenols
3400–3250 (m)	N–H	Str	Primary and secondary amines, amides
3300–2500 (m)	O–H	Str	Carboxylic acids
3330–3270 (n, s)	–C≡C–H:C–H	Str	Alkynes (terminal)
3100–3000 (s)	C–H	Str	Aromatics
3100–3000 (m)	=C–H	Str	Alkenes
3000–2850 (m)	C–H	Str	Alkenes
2830–2695 (m)	H–C=O:C–H	Str	Aldehydes
2260–2210 (v)	C≡N	Str	Nitriles
2260–2100 (w)	–C≡C–	Str	Alkynes
1760–1665 (s)	C=O	Str	Carbonyls (gen)
1760–1690 (s)	C=O	Str	Carboxylic acids
1750–1735 (s)	C=O	Str	Esters, satd aliphatic
1740–1720 (s)	C=O	Str	Aldehydes, satd. aliphatic
1730–1715 (s)	C=O	Str	α, β-unsaturated esters
1715 (s)	C=O	Str	Ketones, satd aliphatic
1710–1665 (s)	C=O	Str	α, β-unsaturated ketones
1680–1640 (m)	–C=C–	Str	Alkenes
1650–1580 (m)	N–H	Bend	Primary amines
1600–1585 (m)	C–C	Str(in-ring)	Aromatic
1550–1475 (m)	N–O	Asym. str	Nitro compounds
1470–1450 (m)	C–H	Bend	Alkanes
1370–1350 (m)	C–H	Rock	Alkanes
1360–1290 (m)	N–O	Sym. str	Nitro compounds
1335–1000 (s)	C–O	Str	Alcohols, carboxylic acids, esters, ethers
1300–1150 (m)	C–H	Wag(–CH$_2$X)	Alkyl halides
1250–1020 (m)	C–N	Str	Aliphatic amines
1000–650 (s)	=C–H	Bend	Alkenes
950–910 (m)	O–H	Bend	Carboxylic acids
910–665 (s, b)	N–H	Wag	Primary and secondary amines
900–675 (s)	C–H "oop"	Aromatics	
850–550 (m)	C–Cl	Str	Alkyl halides
725–720 (m)	C–H	Rock	Alkanes
700–610 (b, s)	–C≡C–H:C–H bend	Alkynes	
690–515 (m)	C–Br	Str	Alkyl halides

s, strong; sh, sharp; m, medium; n, narrow; w, weak; v, variable.

3.6 IR SPECTRA AND COMPLEXES

Infrared (IR) spectroscopy [2, 3] helps to determine the functional groups in organic molecules. This casts doubt as to how this will be useful in the study of metal complexes. There are several methods to detect complex formation in solution, such as change in conductance, change in color, etc. However, confirmation of complex formation is obtained from the IR spectrum. First, the IR spectrum of the pure ligand is recorded and then that of the complex is recorded. Both spectra are then compared. If there are changes in the positions of the bands due to stretching or bending vibrations of the ligand in the spectrum of the complex, then it is confirmed that the ligand has coordinated to the metal and the complex has been formed. In addition to this change in the positions of the bands, a new band due to metal-atom stretching will appear. As an example, if an amine coordinates to a metal through nitrogen, then there will be a change in the positions of the bands due to the stretching frequency of the N–H bond and C–N bond. In addition, a new band due to metal-nitrogen stretching will appear in the spectrum of the complex. This confirms that the amine ligand has coordinated to the metal ion.

Hence, it becomes necessary for us to know the positions of bands for different groups, which are given in Table 3.6.

3.7 COORDINATION AND LIGAND VIBRATIONS

3.7.1 Nitrato Ligand, NO_3^-

The resonance structure of nitrate ion is shown in Fig. 3.12. There are four ways in which this nitrate ion can coordinate to a metal, at least theoretically, as shown in Figs. 3.13–3.16. Of these four possibilities, the symmetrical bidentate chelate structure (Fig. 3.14) is preferred as it is more stable. The free nitrate ion (Fig. 3.12) has D_{3h} symmetry. This symmetry is comparatively high and its IR spectrum is fairly simple. On coordination to a metal the symmetry is lowered. The symmetry of the symmetrical bidentate chelate (Fig. 3.14) is C_{2v}. As we have seen already, the effect of lowering the symmetry is splitting of degeneracies of vibrations and introducing additional bands in the IR spectrum. In this manner, the free and coordinated nitrate ions can be distinguished and this also serves to confirm the coordination of the ligand to the metal. During coordination, the oxygen atom is pulled by both nitrogen and the metal ion. This causes stretching of the N–O bond with becomes weaker. Hence, the band due to N–O vibration

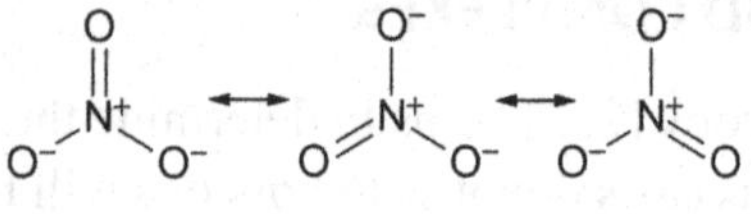

Figure 3.12 Resonance structures of nitrato ligand.

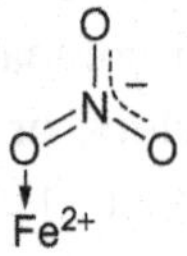

Figure 3.13 Nitrato ligand as monodentate.

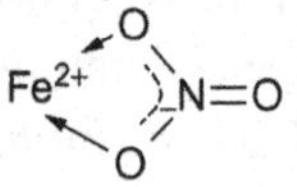

Figure 3.14 Nitrato ligand as symmetrical bidentate chelate.

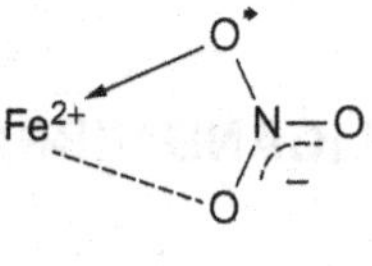

Figure 3.15 Nitrato ligand as unsymmetrical bidentate chelate.

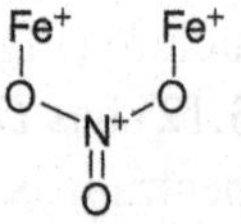

Figure 3.16 Nitrato ligand as bidentate binuclear.

will be shifted to lower energy and hence to lower wavenumber in the spectrum of the complex. In addition to this shift in the position of the band, an additional band due to the metal-oxygen stretching will be seen in the IR spectrum of the complex in the region 470–450 cm^{-1} [4].

3.7.2 Carboxylate Ion

The structure of carboxylate ion is given in Fig. 3.17 and it has C_{2v} symmetry. The free acetate ion has symmetric and antisymmetric C–O stretching modes at approximately 1415 and 1570 cm^{-1} (ν_{COO^-}), respectively. Already the symmetry of the free ion is low (C_{2v}). Hence, further reduction in symmetry due to complexation may not be significant. Therefore, instead of looking for additional bands in the complex, the shift in the positions of bands must be considered as proof of coordination. We can think of at least three ways of coordination for this ligand as shown in Figs. 3.18–3.20. They represent monodentate, symmetrical bidentate chelate, and bidentate binuclear, respectively. When the ligand acts as a monodentate ligand (Fig. 3.18), one C–O bond has double bond character and hence the corresponding stretching band ($\nu_{C=O}$) will appear at a higher wavenumber. This can be used as a tool to identify the monodentate carboxylate ligand. Thus, in

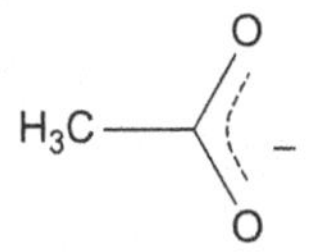

Figure 3.17 Structure of carboxylate ligand.

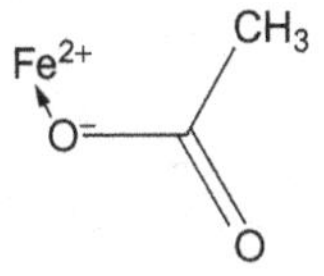

Figure 3.18 Carboxylate as a monodentate ligand.

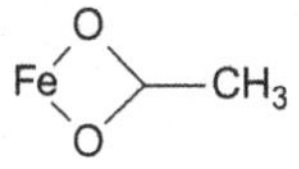

Figure 3.19 Carboxylate as a symmetrical bidentate chelate ligand.

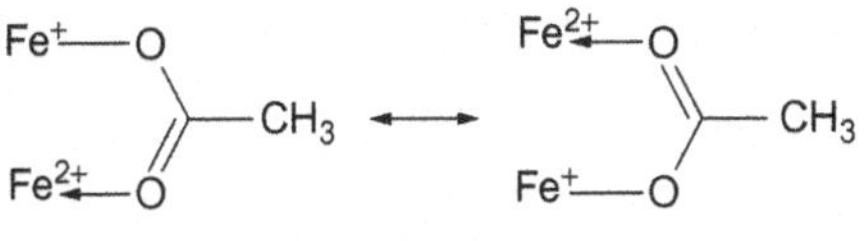

Figure 3.20 Carboxylate as a bidentate binuclear ligand.

the free ligand both the C–O bonds are equivalent and only one band is seen due to (ν_{COO^-}). However, in the complex, where the ligand acts as a monodentate ligand, two bands, one corresponding to ν_{C-O} and the other corresponding to $\nu_{C=O}$, will be seen in addition to the ν_{M-O} band. However, in the case of symmetrical bidentate and bidentate binuclear/bridging, the two C–O bonds are still equivalent and hence only one band due to ν_{C-O} will be seen in addition to the ν_{M-O}. Generally, multiple bands appear between 1400 and 1550 cm^{-1}.

3.7.3 Sulfate Ion

The free sulfate ion has T_d symmetry and its structure is shown in Fig. 3.21.

It can act as a monodentate (Fig. 3.22), bidentate chelate (Fig. 3.23), and bidentate bridging (Fig. 3.24) ligand.

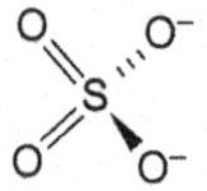

Figure 3.21 Structure of free sulfate ion.

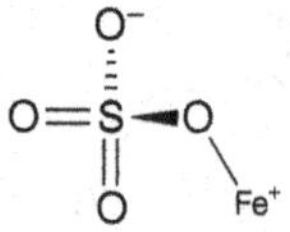

Figure 3.22 Sulfate as a monodentate ligand.

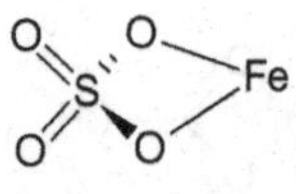

Figure 3.23 Sulfate as a bidentate chelate ligand.

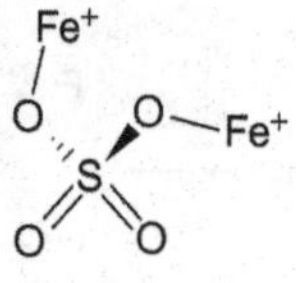

Figure 3.24 Sulfate as a bidentate bridging ligand.

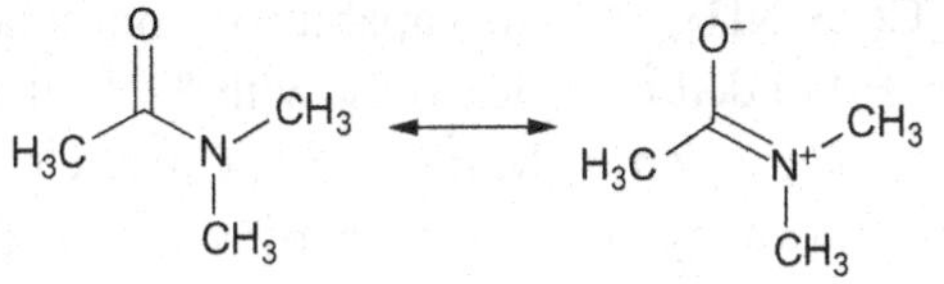

1-(dimethyliminio)ethanolate - *N,N*-dimethylacetamide (1:1)

Figure 3.25 Structure of N, N-dimethylacetamide.

The monodentate (Fig. 3.22) has C_{3v} symmetry, while the bidentate chelate (Fig. 3.23) and bidentate bridging (Fig. 3.24) have C_{2v} symmetries. Thus, we see a reduction in symmetry from the highly symmetrical T_d symmetry. This leads to the appearance of additional bands in the complexes. The free ion shows *one*, monodentate complex shows *three* and bidentate shows *four* v_{S-O}s. However, the bidentate bridging and bidentate chelate cannot be distinguished based on the number of bands because both have the same symmetry, C_{2v}. Hence, they can be distinguished based on the positions of the bands. The bridging ligand has v_{S-O} bands at lower frequencies than the chelating ligand.

3.7.4 *N, N*-Dimethylacetamide

The structure of N,N-dimethylacetamide is shown in Fig. 3.25. Since it contains both nitrogen and oxygen, oxygen coordination and nitrogen coordination are possible and the IR spectrum can help in identifying this. The contributing structure on the left-hand side of Fig. 3.25 is suitable for N-coordination and that on the right-hand side of the figure is suitable for oxygen coordination. Thus, in oxygen coordination, $v_{C=O}$ will disappear, while v_{C-N} will increase as it acquires double bond character and if it is N-coordination, $v_{CC=O}$ will appear at a higher frequency compared to the free ligand because resonance with the nitrogen lone pair is no longer possible.

Actually, the IR spectrum of this ligand in CCl_4 shows a band around 1660 cm^{-1}. This band is due to the carbonyl absorption. But this value is very low compared to the $v_{C=O}$ of acetone (1715 cm^{-1}). This is due to the resonance between the carbonyl π-bond and the nitrogen lone pair.

For example, let us consider urea. It complexes with Fe^{3+}, Cr^{3+}, Zn^{2+}, and Cu^{2+} through oxygen and this is indicated by a decrease in $v_{C=O}$ and increase in v_{CN}. However, urea complexes with Pd and Pt through nitrogen and this is indicated by increase in $v_{C=O}$ and decrease in v_{CN}.

Thiourea, $NH_2C(=S)NH_2$, is also capable of coordinating through N and S. It forms an N-bonded complex only with Ti^{4+}. It forms S-bonded complexes with Pt^{2+}, Pd^{2+}, Zn^{2+}, Mn^{2+}, Fe^{2+}, Co^{2+}, Cu^{1+}, Hg^{2+}, Cd^{2+}, and Pb^{2+}. This is shown by the presence of ν_{M-S} in the region 300–200 cm^{-1}, increase in ν_{CN}, decrease in ν_{CS}, and no appreciable change in ν_{CN}.

3.7.5 Cyano Complexes

The MO diagram of CN is shown in Fig. 3.26. The cyano ligand usually coordinates to metal through carbon. However, isocyano complexes are also possible, where it will coordinate through nitrogen. In some cases both are in equilibrium in the liquid phase. The CN and NC modes of coordination can be easily distinguished. The ν_{CN} appears at higher frequencies in cyano complexes compared to isocyano complexes.

The cyano complexes are easily identified by their sharp ν_{CN} in the range 2200–2000 cm^{-1}. Free CN^- exhibits ν_{CN} at 2080 cm^{-1} in aqueous solution. When this coordinates to a metal, ν_{CN} increases because CN^- is a good σ-donor but a poor π-acceptor. When it donates electrons to a metal, the electrons flow from the 5σ orbital and this is weakly antibonding. Hence, on removal of electrons from this weakly antibonding orbital, the C–N bond order increases and hence ν_{CN} increases.

Factors Affecting ν_{CN}

There are three factors which affect ν_{CN}:

1. electronegativity
2. oxidation state and
3. coordination number of the metal.

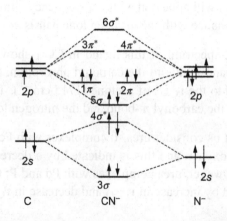

Figure 3.26 MO diagram of CN.

Effect of Electronegativity

Let us consider the three complexes whose ν_{CN} values are given in Table 3.7.

Nickel has the lowest electronegativity compared to the other two. That is, its capacity to attract electrons is less. Hence, σ-donation from CN^- decreases. In turn, electron flows less readily from the 5σ orbital, which is weakly antibonding as stated earlier. Therefore, the C–N bond is weaker in this complex compared to the other two. Thus, it has the lowest frequency for ν_{CN}.

Effect of Oxidation State

When the oxidation state of the central metal ion increases, it readily accepts electrons from ligands, and here, the σ-donation from CN^- increases. Since these electrons flow from the weakly antibonding orbital, the C–N bond strength increases and hence ν_{CN} increases. This is shown by the example given in Table 3.8.

Effect of Coordination Number

When coordination number increases, that is, when more ligands are coordinated to the central metal ion, the electron density on the metal increases and hence the metal is not able to accept electrons readily from CN^-. Thus, electron flow from the weakly antibonding orbital of CN decreases. This weakens the C–N bond and hence ν_{CN} decreases. In short,

Table 3.7 CN Stretching Frequencies and Electronegativity	
Complex	ν_{CN} (cm^{-1})
$Ni(CN)_4^{2-}$	2130
$Pd(CN)_4^{2-}$	2145
$Pt(CN)_4^{2-}$	2150

Table 3.8 CN Stretching Frequencies and Oxidation State	
Complex	ν_{CN} (cm^{-1})
$V(CN)_6^{5-}$	1912
$V(CN)_6^{4-}$	2060
$V(CN)_6^{3-}$	2080

when the coordination number increases, ν_{CN} decreases. This is evident from Table 3.9.

3.7.6 Dimethyl Sulfoxide Complexes

In the case of this ligand, coordination can occur through sulfur or oxygen. The resonance structure of the ligand is shown in Fig. 3.27. The contributing structure on the left-hand side will contribute to "S" coordination and that on the right-hand side to "O" coordination. Thus, in S-coordination, the S–O bond will have double bond character and in O-coordination it will have single bond character. Hence, in S-coordination, ν_{SO} will have a higher value and in O-coordination ν_{SO} will have a lower value. In free dimethyl sulfoxide (DMSO), ν_{SO} appears around 1100–1055 cm^{-1} and in the cobalt complex, Co(DMSO)$_6$, it appears around 950 cm^{-1}. Since ν_{SO} has decreased, it indicates oxygen coordination. On the other hand, in PdCl$_2$(DMSO)$_2$ and PtCl$_2$(DMSO)$_2$, ν_{SO} has increased and is observed in the region 1158–1118 cm^{-1}, showing sulfur coordination.

3.7.7 Metal Carbonyls

The MO diagram of carbon monoxide is shown in Fig. 3.28. The ligand CO is σ-donating as well as π-accepting. During σ-donation, electrons flow from the 5σ bond, which is slightly antibonding. Hence, CO bond strength increases and hence ν_{CO} also increases. At the same time this CO ligand accepts electrons from the metal into its empty π^* orbitals. Since these are antibonding orbitals, this weakens the CO bond and hence ν_{CO} decreases.

Table 3.9 CN Stretching Frequencies and Coordination Number	
Complex	ν_{CN} (cm^{-1})
Ag(CN)$_4$$^{3-}$	2090
Ag(CN)$_3$$^{2-}$	2103
Ag(CN)$_2$$^-$	2130

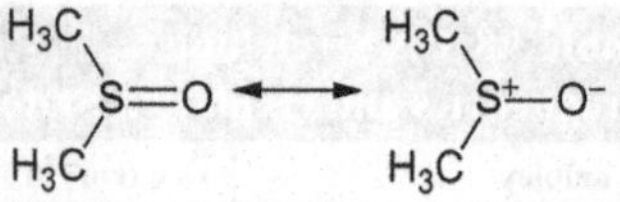

Figure 3.27 Resonance structure of dimethyl sulfoxide.

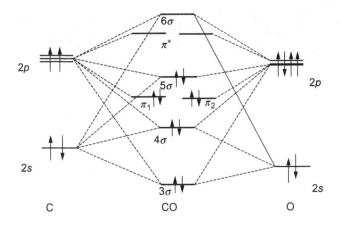

Figure 3.28 MO diagram of carbon monoxide ligand.

Thus, both the effects are synergistic. Since CO is strongly π-bonding, the net effect is the weakening of the CO bond and hence ν_{CO} decreases. ν_{CO} of free CO is in the range 2100–1800 cm^{-1}. This is lowered in complexes, especially when the metal is in a low oxidation state. However, the opposite trend is observed in metal halides because metals are in relatively higher oxidation states.

Terminal and Bridging Carbonyls
When CO acts as a bridging ligand between two metals, ν_{CO} appears around 1900–1800 cm^{-1}, while when CO is a terminal ligand ν_{CO} appears at higher frequency, 2100–2000 cm^{-1}.

Structure of Metal Carbonyls
Mononuclear Carbonyls
Mononuclear carbonyls have only one central metal atom and can have *three* different structures, namely T_d, D_{3h}, and O_h. The structures are shown in Figs. 3.29A-C, respectively.

When the negative charge on the metal increases, back donation is more favored, reducing the ν_{CO}. As an example, when we consider $Fe(CO)_4^{2-}$, $Co(CO)_4^-$, and $Ni(CO)_4$, ν_{CO} decreases when we go from Ni $\rightarrow$ Co $\rightarrow$ Fe. This is in agreement with the amount of negative charge on the metal.

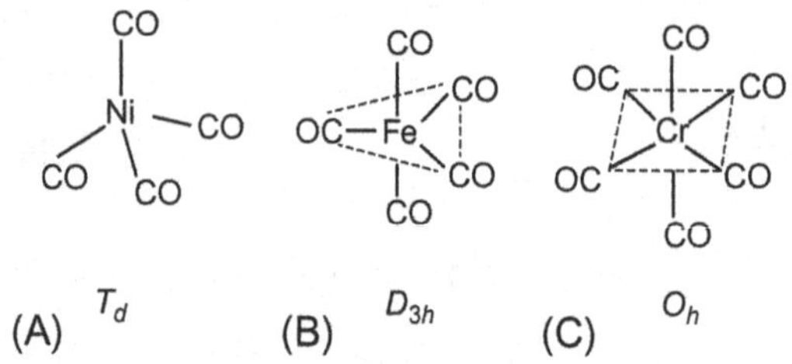

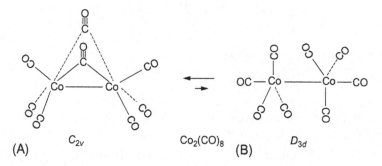

Figure 3.29 Different structures of metal carbonyls. (A) T_d. (B) D_3h. (C) O_h.

Figure 3.30 Structure of $Co_2(CO)_8$. (A) C_{2v}. (B) D_{3d}.

Polynuclear Carbonyls

The structure of $Co_2(CO)_8$ is shown in Fig. 3.30 and that of $Fe_2(CO)_9$ is shown in Fig. 3.31. XRD shows that $Co_2(CO)_8$ takes the structure shown on the left-hand side of Fig. 3.30, having C_{2v} symmetry. It has *six* terminal and *two* bridging carbonyl groups and all are expected to be IR active; ν_{CO} of these terminal groups appears in the range 2080–2025 cm^{-1}. ν_{CO} of the bridging groups appears in the range 1870–1860 cm^{-1}. XRD confirms the structure of $Fe_2(CO)_9$ shown in Fig. 3.31. The structure of $Mn_2(CO)_{10}$ is shown in Fig. 3.32 and has been suggested by XRD.

3.8 ISOTOPIC SUBSTITUTION AND APPLICATION

The frequencies of vibrations very much depend up on the reduced mass of the group and hence on the isotopic mass of atoms. The reduced mass is given by:

$$\mu = \frac{M_A M_B}{(M_A + M_B)} \tag{3.7}$$

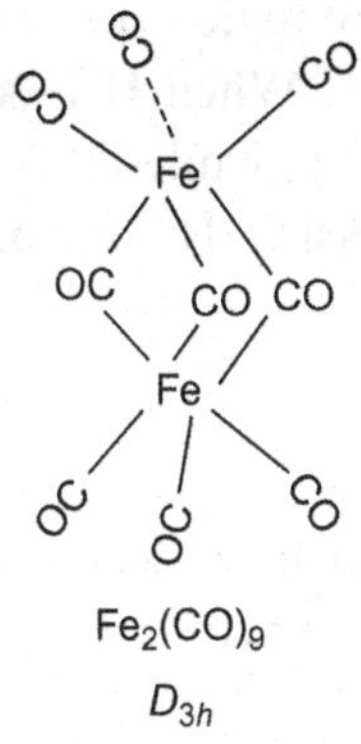

Figure 3.31 Structure of $Fe_2(CO)_9$.

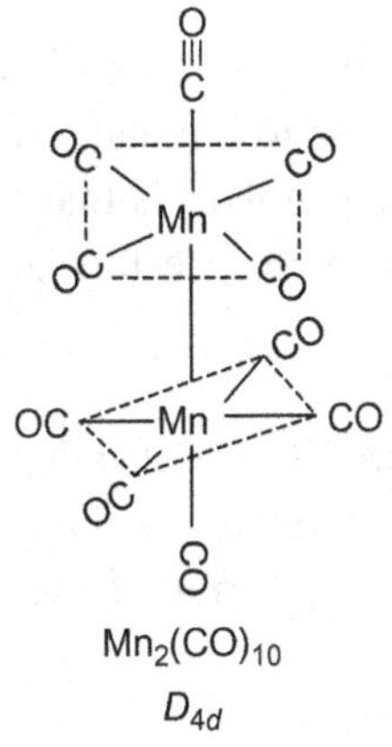

Figure 3.32 Structure of $Mn_2(CO)_{10}$.

where μ is the reduced mass of the system and M_A and M_B are the masses of atoms A and B. Frequency is related to the reduced mass of the system as:

$$\nu = \frac{1}{2\pi}\sqrt{\frac{k}{\mu}} \tag{3.8}$$

where ν is the stretching frequency of the bond and k is the force constant of the bond under consideration. From Eq. (3.8), it is clear that the stretching frequency of a bond (ν) decreases when the reduced mass (μ) increases. Thus, when a heavier isotope is substituted, the stretching frequency of the bond will decrease. This will help us if we want to distinguish two bonds appearing in the same frequency region. For example, ν_{CH} and ν_{CN} appear in the region 3300–2850 cm^{-1}. When H is substituted by D in CH, the frequency is lowered significantly, while that of CN is not disturbed. Thus, it can be confirmed whether it is a C–H bond or CN bond.

EXERCISES

Exercise 3.1. What is meant by reduced mass? Calculate the reduced mass for HCl and DCl, where D = ^{2}H.

Exercise 3.2. An inorganic compound, ML$_3$, is in your laboratory. You want to know whether it has a pyramidal or trigonal planar structure. How will you use IR spectroscopy to solve this?

Exercise 3.3. In a ligand, there is an O–H along with an NH$_2$ group. Even though both can coordinate to a metal, let us assume that only OH is coordinating to the metal. The problem is that the O–H and N–H stretches lie very close. How will you overcome this problem with the help of IR spectroscopy?

Exercise 3.4. Consider the CO$_2$ molecule. How many vibrations are possible for this molecule? How many will be IR active and how many will be Raman active? Give the symmetries of these vibrations and a detailed explanation for your answers.

Exercise 3.5. While in NMR spectra we interpret all the signals, we do not in IR spectra. Why? If the fundamental vibration of a group and first overtone of some other vibration lie very close, can you find which is the fundamental and which is the overtone? If yes, explain.

Exercise 3.6. A complex M(CO)$_5$$^+$ can have structures with either C_{4v} or D_{3h} symmetry. Draw the corresponding structures. How will you determine which is the correct structure out of the two.

Exercise 3.7. What are the possible modes of coordination of thiourea? How will you find the actual coordination?

Exercise 3.8. Explain the following observation: V(CO)$_6$$^-$ shows ν_{CO} around 1850 cm^{-1}, while Cr(CO)$_6$ shows ν_{CO} around 2010 cm^{-1}.

REFERENCES

[1] C.N. Banwell, E.M. McCash, Fundamentals of Molecular Spectroscopy, Tata McGraw Hill, New York, 1994.

[2] W. Kemp, Organic Spectroscopy, Palgrave, New York, 2011.

[3] R.M. Silverstein, F.X. Webster, D.J. Kiemle, Spectrometric Identification of Organic Compounds, John Wiley, New Delhi, 2005.

[4] K. Nakamoto, Infrared and Raman Spectra of Inorganic and Coordination Compounds, Applications in Coordination, Organometallic, and Bioinorganic Chemistry, sixth ed., John Wiley, Hoboken, 2009.

CHAPTER 4

EPR Spectroscopy

This is nothing but electron paramagnetic resonance (EPR) spectroscopy. This is also known as electron spin resonance (ESR). The name itself suggests that the complex to be studied by this technique should be paramagnetic. That is, it should have unpaired electrons. This technique is studied in the microwave region, which has a wavelength ranging from 0.1 to 100 cm. Much valuable information such as the nature of the metal-ligand bond, the geometry of the complex, extent of delocalization of metal d-electrons between metal and ligand, etc., can be obtained using this technique.

4.1 PRINCIPLE OF EPR SPECTROSCOPY

When a paramagnetic molecule is irradiated with microwave frequency, the unpaired electron absorbs that energy and undergoes spin inversion from $-1/2$ to $+1/2$. A signal is produced corresponding to this absorption [1]. In other words, the principle is the same as that of NMR but the energy levels are reversed as shown in Fig. 4.1, where m_s is the magnetic spin momentum quantum number and m_I is the nuclear magnetic spin quantum number, β is the Bohr magneton ($\beta = \frac{eh}{4\pi m_e} = 9.274 \times 10^{-24}\,\mathrm{JT^{-1}}$), H is the applied magnetic field, γ is the magnetogyric ratio, which is specific for a given nucleus ($\gamma = \frac{g\beta_N H}{h}$), β_N is the nuclear magneton ($\beta_N = \frac{eh}{4\pi m_p}$) and g is not a constant and is the spectroscopic splitting factor known as the Landé g-factor. When an electron is not present in any orbital but is free, its g value is 2.00232. Most organic radicals have g values between 1.99 and 2.01, while transition metal species have a wider range of g values.

The spin angular momentum, m_S, of an electron can have two values, namely $+1/2$ and $-1/2$. In the absence of an external magnetic field, these two levels are degenerate. When a magnetic field is applied, the degeneracy is removed and the two levels split. The lower energy level corresponds to $m_S = -1/2$ and the higher level to $m_S = +1/2$. When energy equal to the separation between these two levels is absorbed, transition takes place. This energy falls in the microwave region.

Spectral Methods in Transition Metal Complexes. http://dx.doi.org/10.1016/B978-0-12-809591-1.00004-9

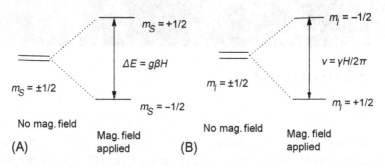

Figure 4.1 Splitting of levels in EPR spectroscopy. (A) EPR. (B) NMR.

4.1.1 Derivative Curves

The signal obtained in EPR is usually a derivative curve. It is obtained by plotting dA/dH versus H, where A is the absorption and H is the magnetic field as shown in Fig. 4.2.

4.1.2 Fine Splitting

Zero-field splitting removes the degeneracy and hence the absorption band is split. This kind of splitting is called *fine splitting*. The lines thus obtained do not have the same intensity but have varying intensities. The central line has the greatest intensity and the outermost lines have the lowest.

4.1.3 Hyperfine Splitting

The spinning electron produces a magnetic field. Similarly, a spinning nucleus also produces a magnetic field. When these two magnetic fields

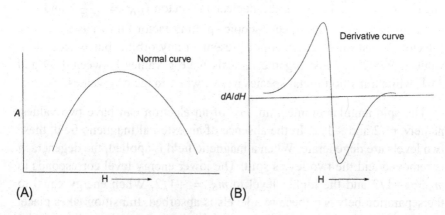

Figure 4.2 (A) Normal curve. (B) Derivative curve in EPR.

interact, the energy levels are split. In other words, when the unpaired electron interacts with a nucleus, splitting of the energy levels will take place. This is called hyperfine splitting.

Explanation

The electron-spin and nuclear-spin coupling arises due to the Fermi contact interaction. This is shown in Fig. 4.3. The Fermi contact interaction is the magnetic interaction between an electron and an atomic nucleus when the electron is inside that nucleus. This combination, namely $m_S = -1/2$ and $m_I = +1/2$ has low energy.

Energy of Levels

The energies of different levels are given by

$$E = g\beta H m_S + A m_S m_I \tag{4.1}$$

where A is the hyperfine splitting constant.

Characteristics of A

1. The sign of A gives an idea about the energy levels: if A is positive, it indicates that the combination, namely $m_S = -1/2$ and $m_I = +1/2$ has lower energy. If A is negative, it indicates that the combination $m_S = +1/2$ and $m_I = -1/2$ has lower energy.
2. The sign of A cannot be determined from a simple spectrum.
3. The magnitude of splitting is expressed in terms of A.
4. The magnitude of A depends on the following:
   (a) It depends on the ratio of the nuclear magnetic moment to the nuclear spin.
   (b) The electron spin density in the immediate vicinity of the nucleus.
   (c) The anisotropic effect.

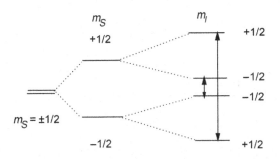

Figure 4.3 Hyperfine splitting in EPR. $m_S = -1/2$ is aligned with the magnetic field and has lower energy. $m_S = +1/2$ is opposed to the magnetic field and has got higher energy.

Selection Rules

The transitions should obey the following conditions:

$$\Delta m_I = 0 \quad \text{and} \quad \Delta m_S = \pm 1$$

How Many Lines?

The number of lines obtained due to hyperfine splitting depends on the value of the nuclear spin quantum number I and the number of lines will be $(2nI + 1)$, where n is the number of nuclei interacting with the unpaired electron. I for some important nuclei are given in Table 4.1.

For example, the EPR spectrum of a copper(II) complex will show four lines due to hyperfine splitting because $I = 3/2$ for Cu(II) as shown in Fig. 4.4.

Table 4.1 Nuclear Spin Values for Some Nuclei		
Nucleus	**Abundance (%)**	**I**
^{7}Li	92.57	3/2
^{45}Sc	100	7/2
^{51}V	99.76	7/2
^{53}Cr	9.54	3/2
^{55}Mn	100	5/2
^{57}Fe	2.245	1/2
^{59}Co	100	7/2
^{61}Ni	1.25	3/2
^{63}Cu	69.09	3/2
^{65}Cu	30.91	3/2

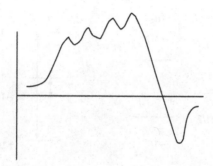

Figure 4.4 EPR spectrum of copper complex.

Hyperfine Splittings in Various Structures

- When an unpaired electron interacts with a nuclear spin I, $2I+1$ lines will be obtained with equal intensity and of equal spacing. For example, if the unpaired electron interacts with N ($I = 1$), three lines will be obtained. In general, when there are n equivalent nuclei of spin I_i, the number of lines obtained will be equal to $2nI_i + 1$.
- When there are two sets of *nonequivalent* nuclei, one set of n nuclei with spin I_i and another set of m nuclei with spin I_j, then the number of lines obtained will be equal to $(2nI_i + 1)(2mI_j + 1)$.
- If the unpaired electron interacts with n *nonequivalent* protons, it will give rise to 2^n lines corresponding to the two spin states of the protons.
- If the unpaired electron interacts with n *equivalent* protons, it will give rise to $(2nI + 1)$ lines.

4.2 g-VALUES IN DIFFERENT ENVIRONMENTS

4.2.1 Solid-State and Frozen Solutions

The number of peaks in EPR depends on the molecular symmetry of the compound. For example, if the molecule contains an axis of symmetry, the electric field in this direction will be different from the remaining two directions.

$g_\perp$ and $g_\parallel$

When the external magnetic field is parallel to the axis of symmetry, the g value obtained is called $g_\parallel$ and when the applied magnetic field is perpendicular to the axis of symmetry it is called $g_\perp$. These are illustrated in Fig. 4.5.

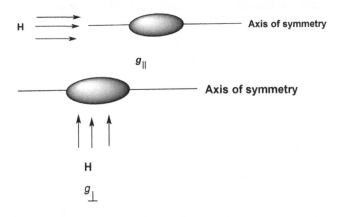

Figure 4.5 $g_\parallel$ and $g_\perp$.

g_{xx}, g_{yy}, and g_{zz}

If the molecule has no axis of symmetry, then all the three directions will be unique and three different g-values will be obtained. These are known as g_{xx} along the x-direction, g_{yy} along the y-direction, and g_{zz} along the z-direction. These are shown in Fig. 4.6.

4.2.2 Characteristics of g

1. It is not a constant.
2. This is a tensor quantity. A tensor has *three* coordinates and *nine* components. A *vector* has *two* coordinates (magnitude and direction) and *three* components. For example, stress is a tensor quantity. It depends on the direction, the magnitude, and the plane under consideration. Similarly, g depends on *three* quantities, namely the electron spin, the orbital angular momentum, and the direction of the external magnetic field.
3. $g = 2.0023$ for a free electron.
4. The g value for paramagnetic metal ions is very different from that of free electrons.
5. The magnitude of g depends on the orientation of the paramagnetic molecule with respect to the external magnetic field.
6. In solution or gas phase, the g value of the paramagnetic molecule is averaged over different orientations and the mean value is obtained. It is called *isotropic*. However, in a crystal, the molecular motion is frozen and different values are obtained for g for different directions and is called anisotropic.
7. If the paramagnetic molecule or ion is located in a crystal of high symmetry such as a cubic crystal, the site will be O_h or T_d, so the g value is independent of direction and is called *isotropic*.

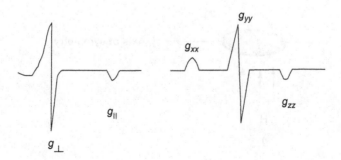

Figure 4.6 $g_{\parallel}$, $g_{\perp}$, g_{xx}, g_{yy}, and g_{zz}.

8. If the crystal is of lower symmetry, then the g value depends on the direction and is said to be *anisotropic*.

9. The three directions are defined as follows. The z-direction is defined to be coincident with the highest-fold rotation axis (ie, principal axis), which can be determined by XRD. The g value determined when the z-axis is parallel to the magnetic field is called g_z and is also known as $g_{\parallel}$.

10. Values of g along the x- and y-axes are called g_x and g_y. In a tetragonal field, g_x and g_y are equal and are collectively known as $g_{\perp}$.

11. When the molecule is oriented in some other direction, let θ be the angle between the magnetic field direction and the z-axis of the molecule, then the value of g is obtained from:

$$g^2 = g_{\parallel}^2 \cos^2 \theta + g_{\perp}^2 \sin^2 \theta \qquad (4.2)$$

12. The inequality in g-values helps us to determine the small distortions, if any, in the molecules, which may not be detected by XRD.

Factors Affecting the Magnitudes of g-Values

1. When an unpaired electron is present in a chemical environment or transition metal complex, the g-value will be different from the free electron value.

   Explanation. Two factors, namely quenching and the sustaining effect of *orbital degeneracy*, need to be considered. Quenching roughly means *removing*. Hence, quenching orbital degeneracy means that the degeneracy is removed. This happens when the unpaired electron is placed in a chemical environment or transition metal complex. Jahn-Teller distortion also removes the degeneracy of orbitals.

   Now, we can see what is meant by a sustaining effect. This tries to retain the orbital degeneracy. It prevents the complete removal of orbital degeneracy. This sustaining effect is due to the existence of spin-orbit coupling. Spin-orbit coupling prevents the complete removal of orbital degeneracy but the higher degree of orbital degeneracy is reduced to a lower degree of degeneracy. For example, if there is threefold orbital degeneracy, it may be reduced to twofold degeneracy and so on.

   Spin-orbit coupling means that if an electron possesses orbital angular momentum, it will be retained by coupling to the spin angular momentum and if the electron has spin angular momentum, it will generate orbital angular momentum.

   Origin of Orbital Angular Momentum. If an orbital can be converted to another similar orbital by rotating the orbital about an axis, orbital angular momentum will be present. For example, a d_{xy} orbital can be converted to a d_{yz} or d_{zx} orbital by rotating about the C_4 axis. This is possible because these orbitals have similar shape. However, a $d_{x^2-y^2}$ orbital cannot be converted into a d_{z^2} orbital because they have different shapes. Hence, if an electron is present in a t_{2g} orbital, it will possess orbital angular momentum, whereas an electron present in an e_g orbital will not have orbital angular momentum.

Consequence. Because of these quenching and sustaining effects, orbital degeneracy is not completely removed but only partially removed. Hence, there will be a net orbital magnetic moment for the electron and hence the g-value will be different from the free electron value of 2.0023. In the case of a free electron, orbital degeneracy will be completely removed.

2. *Transition Metal Ions.* In transition metals two important factors have to be considered, namely crystal field splitting and spin-orbit coupling. These have opposite effects. While the crystal field effect removes the degeneracy, spin-orbit coupling tries to keep the degeneracy. Hence, the net effect depends on the relative strengths of these two factors. Based on this, we can have *three* different cases.

(a) Spin-orbit coupling is very much stronger than the crystal field.

   This is the case in rare-earth ions. These have *f-electrons* and these are very well shielded from the crystal field effects because these are deeply buried. As a result, the orbital degeneracies are not removed and the spin-orbit coupling is retained to a greater extent. This gives rise to a large orbital contribution to the magnetic moment and large anisotropic moments arise. As a result, these behave like free electrons. The g-value is very much closer to that of free electrons. The g-value is calculated by

$$g = 1 + \frac{J(J+1) + S(S+1) - L(L+1)}{2J(J+1)} \tag{4.3}$$

   The g-values calculated in this way agree with the experimental values.

(b) The crystal field is strong enough to just break the spin-orbit coupling.

   Here, the spin-orbit coupling is removed to some extent but not completely by the strong crystal fields. Hence, the magnetic moment will be nearer to the spin-only value. The magnetic moment is calculated by the spin-only formula as shown in Eq. (4.4) and the 3d transition metals belong to this category.

$$\mu_S = g\sqrt{S(S+1)} \qquad (4.4)$$

The g-value will be different from the free electron value because the spin-orbit coupling is not completely removed and hence there will be some orbital contribution. This will be clearer when we see the cases of Fe(III) and Mn(II). These are d^5 systems and their ground state will be ^6S. The ground state (S) is orbitally nondegenerate and hence there will be no orbital contribution to the magnetic moment and g-value. Therefore, the g-values in these cases will be very nearly equal to the free electron value.

(c) The crystal field effect is very much greater so that the spin-orbit coupling is broken down completely. This refers to the strong-field case. "Strong-field" refers to the strong crystal field. This applies to the 4d and 5d transition elements and the strong-field case (such as cyano ligands) of 3d transition elements. In these cases, a molecular orbital description gives better results than the crystal field approximation. Example 1: Octahedral nickel(II) complexes. This is a d^8 system. The ground state is $^3A_{2g}$ and the first excited state is $^3T_{2g}$. These two mix and Eq. (4.5) is used to get the g-value:

$$g = 2 - \frac{8\lambda}{10Dq} \qquad (4.5)$$

where λ is the spin-orbit coupling constant. The value of λ can be found via the experimental values of g from EPR spectra and 10Dq from the electronic spectrum of the complex. Let us consider $[Ni(H_2O)_6]^{2+}$. The g-value for this complex is found to be 2.25 and the 10Dq was found to be 8500 cm^{-1}. Substituting these values in Eq. (4.5), we get $\lambda = -270$ cm^{-1}. The free ion value for Ni(II) is -324 cm^{-1}. Thus, the λ-value decreases in the complex. This lowering of value indicates the extent to which the metal and ligand orbitals mix. This example shows how the spin-orbit coupling and Dq affect the magnitude of the g-value.

Example 2: Cu(II) complex in a tetragonal field. The energy levels in different environments are shown in Fig. 4.7. The expressions for the g-values are

$$g_{\parallel} = 2 - \frac{8\lambda}{(E_2 - E_0)} \qquad (4.6)$$

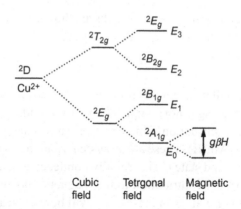

Figure 4.7 Splitting of energy levels of Cu(II) in different fields.

$$g_\perp = 2 - \frac{2\lambda}{(E_3 - E_0)} \qquad (4.7)$$

The values of $(E_2 - E_0)$ and $(E_3 - E_0)$ can be obtained from electronic spectra and g-values from the EPR spectrum. Substituting these values in Eqs. (4.6) and (4.7), we can derive the value of the spin-orbit coupling constant, λ. It can be seen from the equations that the g-value approaches zero when the splitting of the energy levels increases (the differences between E values occur in the denominator).

4.3 ZERO-FIELD SPLITTING AND KRAMER'S DEGENERACY

4.3.1 Zero-Field Splitting

Zero field refers to the absence of a magnetic field. Hence, zero-field splitting refers to the splitting of levels in the absence of an external magnetic field. This happens when more than one unpaired electron is present. The splitting is due to the crystal field.

4.3.2 Kramer's Degeneracy

This rule states that if there is an odd number of unpaired electrons present in the system, each level remains doubly degenerate. This is called Kramer's degeneracy. This is explained as shown in Fig. 4.8.

Example 1: A molecule with *two* unpaired electrons. Total spin $S = 2(-1/2) = -1$. The spin angular momentum m_S has values $-1, 0, +1$.

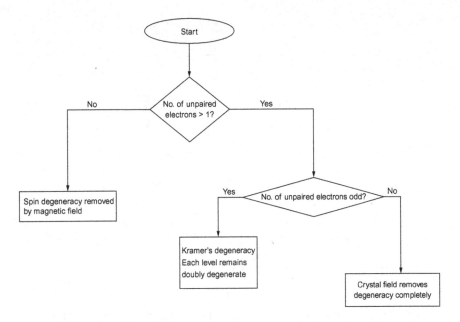

Figure 4.8 Flow chart explaining Kramer's degeneracy.

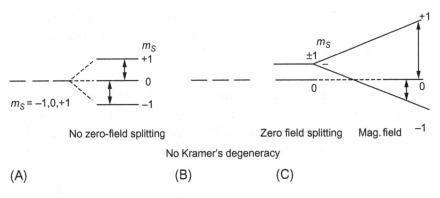

Figure 4.9 Splitting of levels without Kramer's degeneracy.

Since the number of unpaired electrons is even, Kramer's degeneracy will not operate and the crystal field effect removes the degeneracy completely as shown in Fig. 4.9A. The two transitions will have the same energy and hence only one signal will be seen in the absence of a magnetic field (Fig. 4.9B). When a magnetic field is applied further splitting will take place as shown in Fig. 4.9C.

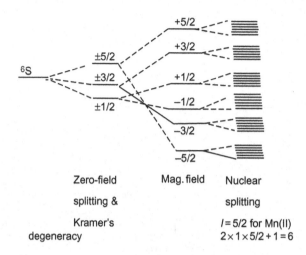

Figure 4.10 Energy levels of Mn(II) under zero-field splitting and Kramer's degeneracy.

Example 2: A molecule with an odd number of unpaired electrons. Mn(II) is a d^5 system containing *five* unpaired electrons. Since the number of unpaired electrons is odd, Kramer's degeneracy operates as explained in Fig. 4.10.

Separation Between Lines and θ   In simple cases, the separation between lines depends on the angleθ between the direction of the magnetic field and the z-axis. This relation is given as $3\cos^2\theta - 1$.

4.3.3 Magnitude of Zero-Field Splitting and Signal

Sometimes, the magnitude of the zero-field splitting will be very high and may exceed the energies of the usual EPR transitions as in the case of V^{3+}. In these cases, lines corresponding to $\Delta m_S = \pm 1$ may not be seen but a weak transition corresponding to $\Delta m_S = 2$, that is, between $m_S = \pm 1$, may be observed and this line will be split into *eight* lines because $I = 7/2$ for ^{51}V.

4.4 EFFECTIVE SPIN, S'

As the name implies, this is not the actual spin of the metal ion but different from that. We will consider *two* examples to understand this concept, namely Ni(II) and Co(II) complexes, because Ni(II) complexes have an A-ground state and Co(II) have T-ground state.

Example 1: Nickel(II) Complexes. Ni(II) has $^3A_{2g}$ ground state and A is an orbital singlet state. Hence, the splitting of this state is negligible or not possible. Hence, the *effective spin* S' is same as the original spin S. When zero-field splitting takes place, $2S$ transitions will be observed. In the case of Ni(II) in an octahedral field, the configuration is $t_{2g}^6 e_g^2$ leaving *two* unpaired electrons and hence $S = 2(1/2) = 1$. Zero-field splitting and effect of magnetic field are shown in Fig. 4.9. This gives rise to *two* transitions as shown in the figure.

Example 2: Co(II) in a cubic field has ^4F as the ground state having an actual spin of 3/2 (three unpaired electrons). However, its EPR spectrum shows one signal suggesting a spin of 1/2. This suggested spin is known as the effective spin (S').

Explanation
The ^4F ground state gives rise to a T state in a cubic field leading to a series of Kramer's doublets because of lowering of symmetry and spin-orbit coupling. However, the lowest doublet is largely separated from the immediate higher doublet. Hence, the EPR shows only one line due to the transition between this lowest doublet. Thus the effective spin is equal to 1/2, that is, $S' = 1/2$ instead of the actual spin $S = 3/2$.

4.5 MIXING OF STATES AND ZERO-FIELD SPLITTING

"Zero-field splitting" means splitting of states in the absence of an external magnetic field. Splitting of states in the absence of a magnetic field takes place due to a crystal field and this crystal field splitting will operate only in the case of an asymmetric electron distribution. Hence, Mn(II), which has a symmetric ^6S ground state, will *not* be split by a crystal field. However, Mn(II) also shows zero-field splitting.

Explanation
The symmetric ground state mixes with the asymmetric excited state by spin-orbit coupling. Hence, Mn(II) also shows zero-field splitting. Spin-spin interaction may also lead to splitting but here it is negligible because of the sextet state. However, in the case of *triplets* spin-spin interaction is solely responsible for the zero-field splitting. In certain symmetries, spin-spin interaction even makes the $\Delta m_S = 2$ transition allowed.

4.6 ANISOTROPY IN HYPERFINE COUPLING CONSTANT

When will the value of g will be isotropic, that is, the same value in all directions? The molecule should be spherically symmetric or should be rotating rapidly in all directions in order to have an *isotropic g*-value. The same conditions hold good for hyperfine coupling constants as well. That is, the molecule must be spherically symmetric or should be rapidly rotating so that the hyperfine coupling constant is *isotropic*. The energy of transition is given by Eq. (4.8). This expression does not consider the fine splitting, that is, the interaction with the metal nucleus.

$$E = h\nu = g\beta H = Am_I \tag{4.8}$$

where A is the isotropic hyperfine coupling constant and m_I is the nuclear spin momentum of the interacting nucleus in the ligand. In the gas phase or in solution, the value of A will be isotropic because of rapid tumbling of the radical or metal complex. However, in the solid matrix or crystal, the molecule is fixed in a particular orientation and hence, the magnetic field will be anisotropic, that is, depends up on the direction. Now, the hyperfine coupling constant will be anisotropic and is represented as B. Hence, the coupling constant can be factored into *isotropic* and *anisotropic* components. The g-value and the A-value have the same angular dependence for anisotropy, namely $3\cos^2\theta - 1$. Hence, Eq. (4.8) can be modified to Eq. (4.9) to consider the anisotropy in both g and A for a system with axial symmetry:

$$E = h\nu = \left(\frac{1}{3}g_\parallel + \frac{2}{3}g_\perp\right)\beta H + Am_I + \left[\frac{1}{3}(g_\parallel - g_\perp)\beta H + Bm_I\right](3\cos^2\theta - 1) \tag{4.9}$$

where θ is the angle between the direction of the magnetic field and the z-axis, "A" is the isotropic hyperfine coupling constant, and B is the anisotropic hyperfine coupling constant.

When the complex or radical rotates freely, $\theta = 0$, and hence Eq. (4.9) reduces to Eq. (4.8). When the sample is powdered, a large number of crystals with random orientation are present. Now, θ has a wide range of angles and hence a broad line is obtained in EPR.

The magnitude of the hyperfine coupling constant usually does *not* depend on the external magnetic field because the nucleus is not directly affected by the external magnetic field.

In the case of metal ions, the coupling constant is expressed as "A" (or $A_\parallel$) and "B" (or $A_\perp$). These have the same significance as $g_\parallel$ and $g_\perp$.

4.7 LINE WIDTHS IN SOLID-STATE EPR

The width of EPR signals depends on the *two* relaxation processes, namely spin-spin relaxation and spin-lattice relaxation.

4.7.1 Spin-Lattice Relaxation

Spin-lattice relaxation leads to line broadening. Here, lattice refers to the surrounding molecules or atoms. When the paramagnetic ion interacts with the thermal vibrations of the lattice, spin-lattice relaxation results. In some compounds this spin-lattice relaxation is sufficiently long so that the EPR spectrum can be observed at room temperature. In some cases it may be too short so that it may not be possible to observe the spectrum at room temperature. The relaxation time is inversely proportional to the temperature. That is, when the temperature is decreased, the relaxation time increases and the spectrum can be recorded. Hence, many transition metal salts and complexes have to be cooled to liquid nitrogen, hydrogen, or helium temperatures to record good EPR spectra.

4.7.2 Spin-Spin Relaxation

The magnetic fields produced by the spin of the paramagnetic ions interact. This is called spin-spin relaxation. Due to this, the total magnetic field at the ions is slightly altered and the energy levels are shifted and distribution of energies takes place. This results in the broadening of signals. This effect depends on two parameters, namely (1) "r," the distance between the paramagnetic ions, and (2) θ, the angle between the field and the symmetry axis; and the relation is $(\frac{1}{r^3})(1 - 3\cos^2\theta)$. This relation indicates that the broadening of signals can be reduced by increasing the distance between the paramagnetic ions and by diluting the salt or the complex with an isomorphous diamagnetic material. For example, small amounts of paramagnetic $CuSO_4$ are mixed in a large amount of diamagnetic $ZnSO_4$ crystals.

4.7.3 Spin Exchange Processes

Exchange of spin between ions can lead to broadening of signals. This can also be reduced by dilution. This exchange is of *two* types: exchange between equivalent ions and between dissimilar ions.

Exchange Between Similar or Equivalent Ions
When the exchange occurs between equivalent ions, the lines broaden at the base and become narrower at the center.

Exchange Between Dissimilar Ions

When the exchange is between dissimilar ions, the lines merge and a broad line is produced. This kind of effect is observed in $CuSO_4 \cdot 7H_2O$ because it has two distinct copper sites per unit cell.

4.8 APPLICATIONS OF EPR

Example 1: Bis-salicylaldiminecopper(II). The structure of this complex is shown in Fig. 4.11.

Cu(II) is a d^9 system having one unpaired electron. $I = 3/2$ for Cu(II) and hence the line is split into *four* lines by the copper nucleus. The ligand contains *two equivalent nitrogens* which has $I = 1$. Hence, each line is split into $(2 \times 2 \times 1 + 1) = 5$ lines. These are further split by the two H′ atoms into $(2 \times 2 \times 1/2 + 1) = 3$ lines. Thus, each of the four lines is split into $5 \times 3 = 15$ lines. However, due to overlap of these lines, only 11 lines are seen. It can be determined whether H′ or H″ atoms are involved in splitting as follows: when both the H″ atoms are deuterated, there was no change in the original spectrum showing that these are not interacting with the unpaired electron. However, when H′ atoms are substituted by CH_3 groups, the splitting due to hydrogens disappears and only four lines due to copper splitting are seen. Thus, four lines due to copper $[(2 \times 3/2 + 1 = 4)]$ are seen, each line being split into five lines due to two nitrogens $[(2 \times 2 \times 1 + 1 = 5)]$.

In this way, the EPR spectrum gives concrete proof of metal-ligand coordination. If a complex were not formed, then the splitting due to nitrogens and hydrogens would not be seen. Only the splitting due to copper alone would be seen in the spectrum. In other words, when the metal electron is delocalized between metal and ligand, a covalent bond is formed between metal and ligand, ie, metal-ligand bond, and this leads to hyperfine splitting of the EPR lines.

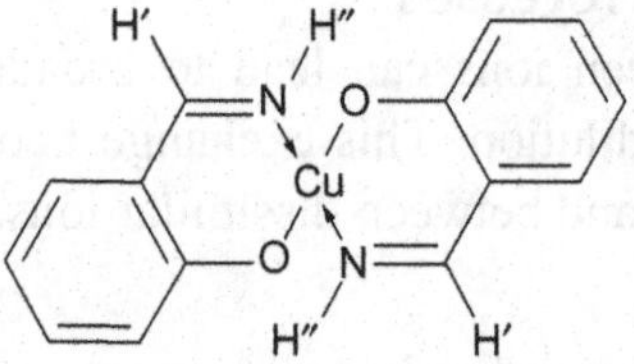

Figure 4.11 Structure of bis-salicylaldiminecopper(II).

Example 2: $CuSiF_6 \cdot 6H_2O$. This complex is diluted with the diamagnetic Zn salt. It was cooled to 90 K and the EPR spectrum was recorded. The spectrum consisted of one band with partially resolved hyperfine structure and a nearly isotropic value of g was obtained. The ground state of Cu(II) has E symmetry and is doubly degenerate. The above complex has trigonal symmetry rather than cubic symmetry. However, the orbital degeneracy is not destroyed. Therefore, Jahn-Teller distortion will occur.

This distortion occurs along all the three axes. All these have the same energy and the orbital degeneracy is removed. Hence, three EPR transitions are expected, one for each distortion. However, only one transition was observed. This indicates that the crystal field resonates among the three directions.

But when the temperature is lowered, the spectrum becomes anisotropic and shows three different lines corresponding to copper ions in three different environments caused by three different tetragonal distortions.

Example 3: $[(NH_3)_5Co-O-O-Co(NH_3)_5]^{5+}$. There are four structures proposed for this complex:

- There may be two cobalt(III) atoms connected by an O_2^- bridge.
- One may be cobalt(III) and the other cobalt(IV) connected by a peroxy, O_2^{2-} bridge.
- Both the cobalt atoms are made equivalent by equal interaction of one unpaired electron with both cobalt atoms.
- The electron interacts unequally with both the cobalt atoms.

Discussion
1. If the first structure were correct, a single EPR line would result.
2. The second structure would give rise to eight lines ($I = 7/2$ for Co).
3. The third structure will result in 15 lines.
4. The fourth structure would give rise to 64 lines.

However, the actual spectrum consists of 15 lines. This eliminates the second and fourth structures but supports structure 3. However, the first structure is not eliminated. When the hyperfine coupling constant due to cobalt in this complex was compared with that for other cobalt complexes, it was found to be low in this complex. Structure 1 removes electron density from cobalt and hence the magnitude of "A" will decrease. Hence, the contribution of structure 1 is appreciable.

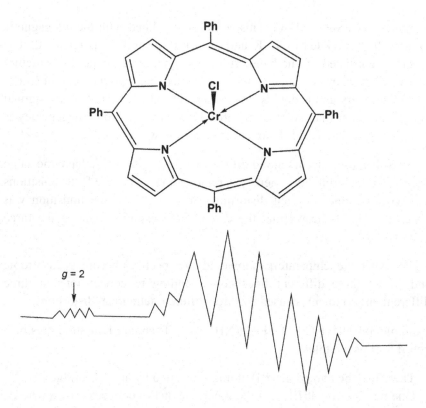

Figure 4.12 Structure of chromium porphyrin derivative and EPR spectrum of the oxidized species.

Example 4: Chromium(III) porphyrin derivative. The structure is shown in Fig. 4.12.

EPR gives direct information and evidence for the number and nature of nuclei with which the unpaired electron is interacting; the magnitude of the hyperfine coupling constant helps to find the extent of delocalization of the unpaired electron(s). The g-values help us to find whether the unpaired electron is based on the metals or ligands. The chromium(III) porphyrin derivative can explain all of this.

The nine intense lines at the center of the spectrum are due to the coupling with the four nitrogens ($2 \times 4 \times 1 + 1 = 9$). In addition there are four satellite peaks on the left-hand side of the spectrum, which are due to the interaction with ^{53}Cr (9.6% abundant and $I = 3/2$, $2 \times 1 \times (3/2) + 1 = 4$).

The fact that peaks are obtained due to interaction with Cr shows that the unpaired electron is centered on the metal in the product obtained. [17]O-labeled experiments also showed that an oxygen atom was bound to Cr in the product. Based on these facts, it was decided that the product was a d^1 oxochromium(V) species formed by a two-electron oxidation reaction.

There may be an alternative product in which the unpaired electron is based on the ligand. It will be a chromium(IV) porphyrin π cation radical. This compound will show some proton hyperfine interactions and the g-value may be very close to 2.002. However, the observed g-value was 1.982. This is consistent with the proposed structure in which the spin-orbit coupling is significant.

In this way, the patterns of hyperfine splittings observed in EPR spectra give direct information about the numbers and nature of spinning nuclei interacting with the unpaired electron(s).

In addition, the magnitude of the hyperfine coupling constants indicates the extent of delocalization of the odd electron(s). The g-values indicate whether the odd electrons are based on the transition metal ions or on the adjacent ligands.

4.9 g-VALUES FOR DIFFERENT GROUND TERMS

T_g Ground Terms

Consider a d^8 system in a cubic field. Let it undergo a tetragonal distortion. In the cubic field, the ground term is $^3A_{2g}$ and the first excited term is $^3T_{2g}$. When it undergoes distortion, the excited $^3T_{2g}$ term undergoes splitting as shown in Fig. 4.13.

The T_{2g} term is split into the nondegenerate $^3B_{2g}$ and doubly degenerate ^3E_g components. The g-value deviates from 2.00 for A_g and E_g ground terms because they interact with the higher T_{2g} terms via spin-orbit coupling. The g-values for A_{2g} and E_g ground terms can be obtained from Eqs. (4.10) and (4.11) [2]. For A_{2g} terms,

$$g_{\parallel} = 2\left(1 - \frac{4k\lambda}{10Dq}\right) \tag{4.10}$$

$$g_{\perp} = 2\left(1 - \frac{k\lambda}{10Dq}\right) \tag{4.11}$$

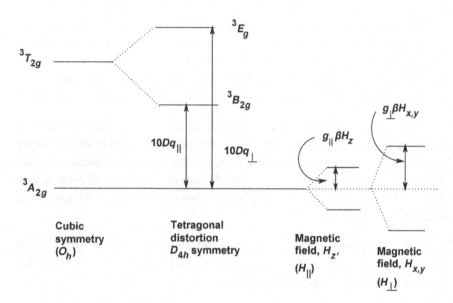

Figure 4.13 Tetragonal distortion and g-value.

where λ is the spin-orbit coupling constant and "k" gives the extent of delocalization of an electron between the metal and the ligand. For E_g terms there is a complication because reduction in symmetry increases the orbital degeneracy and the effect of the magnetic field depends on which orbital component remains as the ground term in the lower symmetry. In either case, the g-value is anisotropic. In tetragonal distortion, ∥ means that the magnetic field is parallel to the C_4 axis and in trigonal distortion it means that the magnetic field is parallel to the C_3 axis. In both these symmetries, $g_\parallel = g_z$ and $g_\perp = g_x = g_y$. The g_{av} is given by

$$g_{av} = \sqrt{\left(\frac{g_\parallel^2}{3} + 2\frac{g_\perp^2}{3}\right)} \tag{4.12}$$

4.10 g-VALUE AND STRUCTURE

Let us consider a copper(II) complex of cubic symmetry. It is a d^9 system, will be the same as an inverted d^1 system and will undergo Jahn-Teller distortion. Let it be tetragonally distorted, that is, elongated or compressed

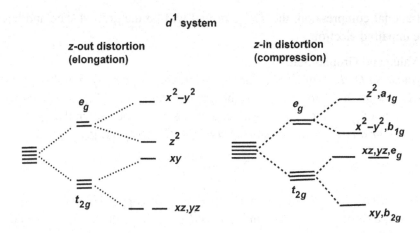

Figure 4.14 Tetragonal distortion in cubic symmetry.

along the C_4 axis coinciding with the z-axis. The splitting of the levels is shown in Fig. 4.14.

For filling the electrons, the normal diagram (Fig. 4.14) should be used and for assigning the transitions the hole formalism is used. For example, in a copper(II) complex, if the ground state is $d_{x^2-y^2}$, it means that the unpaired electron is in the $d_{x^2-y^2}$ orbital. Actually, this orbital has the highest energy (Fig. 4.14). However, it has a hole (d^9 system) and hence it becomes the ground state in terms of hole [1].

Ground State in Elongation and Compression in Tetragonal Distortion

Elongation   When there is elongation along the z-axis (C_4-axis), the orbitals with z-component are lowered in energy and thus the d_{z^2} orbital has lower energy than $d_{x^2-y^2}$. Hence, the unpaired electron will be in the $d_{x^2-y^2}$ orbital. Even though this orbital has the highest energy, this becomes the ground state because it has the hole (Fig. 4.14).

Compression   During compression, those orbitals having the z-component will come closer to the metal orbitals and hence will be repelled and therefore the energies of these orbitals will increase (Fig. 4.14). Thus, the d_{z^2} orbital has higher energy than $d_{x^2-y^2}$. Therefore, the unpaired electron will be in the d_{z^2} orbital. Even though this orbital has the highest energy, it will be the ground state because it has the hole.

In short, during tetragonal elongation, the d_{z^2} orbital will be the ground state and the unpaired electron will reside in that orbital and, during tetragonal compression, the $d_{x^2-y^2}$ orbital will be the ground state and has the unpaired electron.

g-Value and Ground State

Ground State, d_{z^2}. For a tetragonally compressed copper(II) complex, $g_\perp > g_\|(g_{zz} = g_\| = 2)$ indicates that the ground state is d_{z^2}. Eq. (4.13) shows the value of $g_\perp$. Since λ is negative, $g_\perp > 2$. It can be shown that $g_\| = 2$. Hence, *for a tetragonally compressed copper(II) complex*, $\mathbf{g}_{xx} = g_\perp > g_\|$.

$$g_\perp = 2 - \frac{6\lambda}{E_{d_z^2} - E_{d_{xz},d_{yz}}} \tag{4.13}$$

Ground State, $d_{x^2-y^2}$. For a tetragonally elongated copper(II) complex, $g_\| > g_\perp > 2$, indicates that the ground state is $d_{x^2-y^2}$. The value of $g_\|$ is given by Eq. (4.14). Since λ is negative, $g_\| > 2$.

$$g_{zz} = g_\| = 2 - \frac{8\lambda}{E_{d_{x^2-y^2}} - E_{d_{xy}}} \tag{4.14}$$

Eq. (4.15) gives the value of $g_\perp$:

$$g_{xx} = g_\perp = 2 - \frac{2\lambda}{E_{d_{x^2-y^2}} - E_{d_{xz},d_{yz}}} \tag{4.15}$$

From Eqs. (4.14) and (4.15), we get $g_\| > g_\perp > 2$ for a tetragonally elongated copper(II) complex.

C_4-Distortion	g-Value	Ground State (ie, Unpaired e^- in)
Elongation	$g_\| > g_\perp > 2$	$d_{x^2-y^2}$
Compression	$g_\perp > g_\| > 2$	d_{z^2}

4.10.1 Magic Pentagon

The magic pentagon [1] helps us to find the g-value under the influence of spin-orbit coupling. The pentagon is shown in Fig. 4.15.

Let us consider a single electron in a nondegenerate d-orbital. Then the g-value is given by

$$g = g_e \pm \frac{n\lambda}{\Delta E} = 2.0023 \pm \frac{n\lambda}{\Delta E} \tag{4.16}$$

where ΔE is the energy difference between the orbital containing the electron and that with which orbital mixing takes place by the spin-orbital coupling process, λ is the spin-orbit coupling constant and n is an integer. When the electron mixes with an empty orbital a plus sign is used and when the electron mixes with a filled orbital a minus sign is used. Now we will

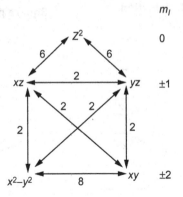

Figure 4.15 Magic pentagon for the calculation of g-values.

see how the magic pentagon is useful. For example, if the transition is taking place from a d_{z^2} orbital to a d_{xz} orbital, $n = 6$ from the magic pentagon. The magic pentagon also gives the m_l values for different orbitals. Now, from Eq. (4.16), we get

$$g_\perp = 2.0023 - \frac{n\lambda}{\Delta E} \quad g_\perp = 2.0023 - \frac{6\lambda}{E_{d_{z^2}} - E_{d_{xz}}}$$

4.11 g-VALUE AND SQUARE PLANAR STRUCTURE

The splitting of d-orbitals in square planar geometry is given in Fig. 4.16.

Let us consider a square planar copper(II) complex. We consider two cases, namely the unpaired electron can be in the $d_{x^2-y^2}$ orbital or in the d_{z^2} orbital. Accordingly, the g-value varies as per Eqs. (4.17)-(4.19). When the unpaired electron is in the $d_{x^2-y^2}$ orbital, the g-values are given by Eqs. (4.17) and (4.18).

Case 1: The unpaired electron is in the $d_{x^2-y^2}$ orbital.

$$g_\parallel = 2.0023 - \frac{8\lambda}{E_{d_{x^2-y^2}} - E_{d_{xy}}} \tag{4.17}$$

$$g_\perp = 2.0023 - \frac{2\lambda}{E_{d_{x^2-y^2}} - E_{d_{xz},d_{yz}}} \tag{4.18}$$

The filling of orbitals and the hole formalism are shown in Fig. 4.17.

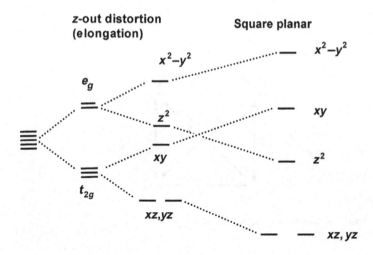

Figure 4.16 Splitting of d-orbitals in square planar geometry.

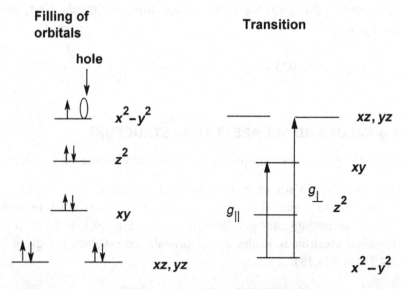

Figure 4.17 Splitting of orbitals and hole formalism for Cu(II).

$g_\parallel > g_\perp > 2$

Case 2: Unpaired electron in the d_{z^2} orbital. The $g_\parallel$ and $g_\perp$ values are given by

$$g_\perp = 2.0023 - \frac{6\lambda}{E_{d_{z^2}} - E_{d_{xz},d_{yz}}} \qquad (4.19)$$

$$g_\parallel = g_e = 2.0023$$

$$g_\perp > g_\parallel = 2$$

4.12 g-VALUE AND COVALENT CHARACTER

In complexes, the metal orbitals containing the unpaired electron(s) mix with the empty or filled ligand orbitals and hence the g-value deviates from the free electron value. The g-value decreases when the metal orbital mixes with the empty ligand orbitals. That is, there is a negative shift $(g - \Delta g)$ in g. Instead, if the metal orbitals mix with the filled ligand orbitals, the g-value increases. That is, there is a positive shift $(g + \Delta g)$ in the g-value [1].

The extent of positive or negative shift depends on the extent of mixing of the metal and ligand orbitals. In other words, the shift in g-value depends on the degree of covalency of the metal-ligand bond. The covalent character and Δg are inversely proportional. That is, the covalent character decreases when the Δg value increases.

When $g_\parallel < 2.3$, the complex is covalent. When $g_\parallel \geq 2.3$, the complex has ionic character.

4.13 $A_\parallel$ AND STRUCTURE

The ratio $g_\parallel/A_\parallel$ is also helpful in determining the structure. For example, when this ratio falls in the range 90–140 cm, it indicates a square planar copper(II) complex [1].

4.14 G-FACTOR AND NATURE OF THE LIGAND

Eq. (4.20) gives the value of G, which gives an idea about the nature of the ligand:

$$G = \frac{g_\parallel - 2.0023}{g_\perp - 2.0023} \tag{4.20}$$

When $G < 4$, it indicates that the ligand is strong and the metal-ligand bond is covalent.

4.15 EPR SPECTRA OF d^n SYSTEMS

The EPR spectra of different paramagnetic d^n systems are discussed in the following sections [3].

4.15.1 d^1 System
Octahedral Field
In an octahedral field, the ground state is $^2T_{2g}$. Since this is a triplet term with degeneracy, an orbital contribution is present. All the Kramer doublets are close in energy and are extensively mixed by spin-orbit coupling and the relaxation time is short leading to broadening of signals.

Tetrahedral Field
The ground state is ^2E containing $d_{x^2-y^2}$ and d_{z^2}. This has no first-order splitting because one orbital cannot be converted into another by any symmetry operation due to their difference in shape. However, the ground state mixes with the $^2T_2$ excited state by second-order spin-orbit coupling. Hence, relaxation time is short leading to signal broadening.

Studying the Spectrum
The EPR spectrum is studied at liquid He temperature. The ^2T excited state splits due to spin-orbit coupling. During distortion, for example, VO^{2+}, the ground state becomes orbitally singlet. The excited states are very much removed from the ground state. Sharp EPR lines are obtained.

4.15.2 d^2 System
Octahedral Field
There is extensive spin-orbit coupling in the $^3T_{1g}$ state. Hence, EPR spectra of only very few complexes have been reported.

Tetrahedral Field
The ground state is $^3A_2$. This is a nondegenerate level and will not have spin-orbit coupling. Hence, relaxation times will be longer and EPR spectra are readily observed.

Examples
V^{3+} octahedral complexes in Al_2O_3 gave $g_\parallel = 1.92$, $g_\perp = 1.63$, and $A = 102$.

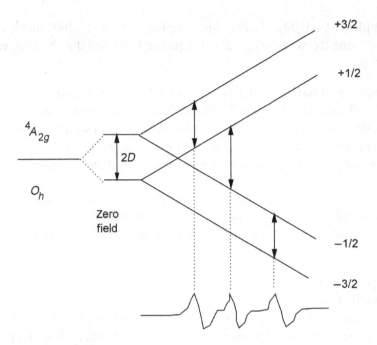

Figure 4.18 Splitting in a d³ system and EPR.

4.15.3 d^3 System
Octahedral Field

These have $^4A_{2g}$ ground state. This will have a lowest energy Kramer doublet. When the zero-field splitting is small as shown in Fig. 4.18, we can expect three transitions. However, when the zero-field splitting is large compared to spectrometer frequency, only one line will be observed.

The g-value for this system is obtained according to:

$$g = 2.0023 - \frac{8\lambda}{\Delta E(^4T_{2g} -^4 A_{2g})} \qquad (4.21)$$

The ground state is $^4A_{2g}$ and it has no spin-orbit coupling because it is orbitally nondegenerate. However, it mixes with the excited $^4T_{2g}$ state to a small extent.

Examples
Example 1: $V(H_2O)_6{}^{3+}$. $\Delta E = 11,750$ cm^{-1} and $\lambda = 55$ cm^{-1}. Substituting these values in Eq. (4.21), we get the value of g as 1.96. This is in agreement with the observed value of 1.97.

Example 2: $Cr(H_2O)_6^{3+}$. For this complex, $\Delta E = 17{,}400$ cm^{-1}, $\lambda = 90$ cm^{-1} and the predicted g-value is equal to 1.961 and the observed value is 1.976.

Example 3: Mn(IV) Complex. Degree of covalency and charge. The calculated g-value is equal to 1.955, while the experimental value is 1.994. This difference is due to the crystal field approximation which becomes less appropriate as the charge increases. Actually, this indicates that covalency becomes more important as the charge on the central ion increases.

> g-value indicates that covalency becomes more important as the charge on the central ion increases (crystal field theory is ionic in nature).

4.15.4 d^4 System
Octahedral Field

The ground state for a weak-field configuration in an O_h field is ^5E_g. As already explained, this has no orbital angular momentum. $S = 4(1/2) = 2$. Possible M_S values are 2, 1, 0, -1, and -2. The splitting of the levels and transitions are shown in Fig. 4.19. When the splitting is small, we can expect four transitions as shown in the figure. However, when the splitting is large, we cannot see any transition. Quite often, we are not able to see any spectrum due to Jahn-Teller distortions and the accompanying large zero-field splitting. Hence, very few spectra are reported for this system.

4.15.5 d^5 System
Strong-Field (Low-Spin) $O_h S = 1/2$

In a strong octahedral field, the ground state is $^2T_{2g}$. This is split into three closely spaced Kramer doublets by spin-orbit coupling. Since the spin-orbit coupling is large, line broadening takes place and hence EPR spectra can be observed only at liquid He temperature. This situation is similar to d^1 but has a positive hole. The expected g-value is not obtained because of distortion, spin-orbit coupling and a magnetic field. The splitting of levels is shown in Fig. 4.20.

Weak-Field, High-Spin Case, $S = 5/2$

Mn(II) complexes have ^6S as the ground state in a weak-field octahedral environment. No other state has this multiplicity of sextet. $^4T_{1g}$ is the nearest excited state. However, mixing with this orbital is possible only if there is strong second-order spin orbit coupling. As a result relaxation time is long

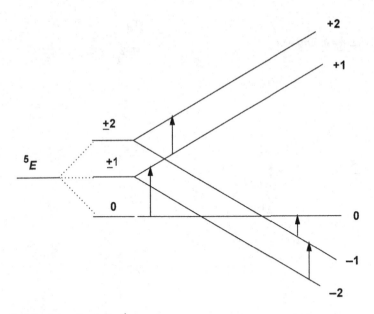

Figure 4.19 EPR splitting of levels in a d^4 system.

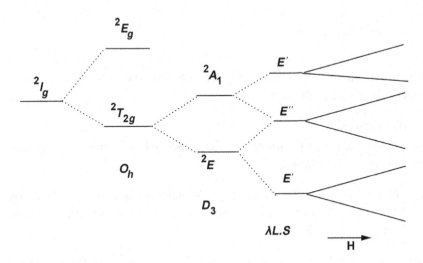

Figure 4.20 EPR splitting of levels in a d^5 strong-field case.

enough to record the EPR spectrum at room temperature. Since there are an odd number of unpaired electrons present (five), Kramer's degeneracy exists even in the presence of strong zero-field splitting. The energy levels and their splitting for Mn(II) are shown in Fig. 4.21. From the figure, it can

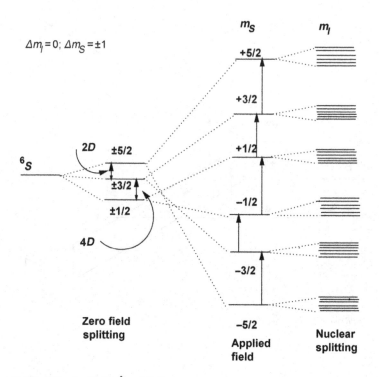

Figure 4.21 EPR splitting in a Mn(II) d^5 weak-field system.

be seen that there will be *five* transitions obeying the EPR selection rules, namely $\Delta m_S = \pm 1$ and $\Delta m_I = 0$.

Fe(III) Complexes

Undistorted Octahedral Complexes   The energy levels and expected spectrum are shown in Fig. 4.22.

Slightly Distorted Fe(III) Complexes   When Fe(III) complexes undergo small tetragonal distortion, the energy levels and expected EPR spectrum are shown in Fig. 4.23.

Highly Distorted Fe(III) Complex   When the tetragonal distortion and zero-field splitting are high, only one EPR line will be observed. The splitting of levels is shown in Fig. 4.24.

4.15.6 d^6 System

Low-spin d^6 complexes are diamagnetic. High-spin O_h complexes are similar to d^4. The high-spin iron(II) complex has a g-value of 3.49 at 4.2 K

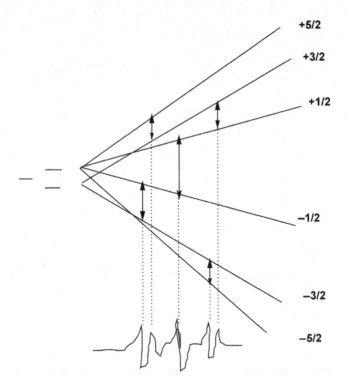

Figure 4.22 EPR splitting of energy levels in undistorted Fe(III).

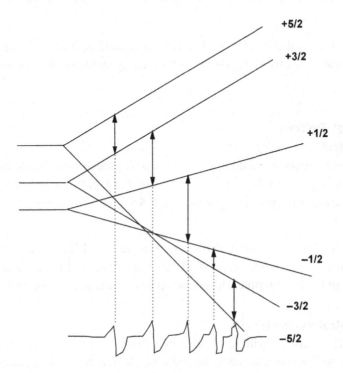

Figure 4.23 EPR splitting of energy levels in distorted Fe(III).

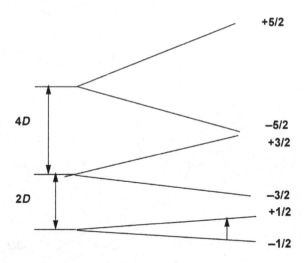

Figure 4.24 EPR splitting in highly distorted Fe(III) complex.

and line width is about 500 gauss even at this low temperature. Spin-orbit coupling is very large even in the ground state and mixes readily with nearby excited states. If the zero-field splitting is small, two transitions will be observed.

When the octahedral field is distorted, zero-field splitting is large and no EPR spectrum is observed. For example, deoxyhemoglobin does not show an EPR.

4.15.7 d^7 System
Octahedral Symmetry

The high-spin octahedral complex has $^4T_{1g}(F)$ ground state. Since the spin-orbit coupling is very high, EPR can be observed only at low temperature. Even at low temperature, only a singlet is seen corresponding to $S' = 1/2$ and $g = 4.33$.

In a strong field, ^2E becomes the ground state, $S = 1/2$. A doublet is seen. Since there is no mixing, relaxation times are long and hence sharp EPR lines are obtained at liquid nitrogen temperature and at room temperature.

Tetrahedral Symmetry

Cobalt(II) complexes with tetrahedral symmetry have $^4A_2$ ground state. This state is closer to the excited state and hence extensive mixing takes place because spin-orbit coupling is strong. The relaxation time is short and hence the signals are broad.

4.15.8 d^8 High-Spin System

The gaseous ion has ^3F ground state and in an octahedral field, the orbital singlet state is the ground state. The d-shell is more than half-filled and hence the g-values are greater than the free electron value. Since the zero-field splitting is high, EPR spectra are not observed at room temperature and can be observed only at low temperatures. The g-values are closer to the isotropic value.

4.15.9 d^9 System

Octahedral Field

The ground state is ^2E_g. Jahn-Teller effect is very strong, splitting the levels to a greater extent. Hence, mixing with higher states is not possible. Hence, EPR spectrum can be observed at room temperature.

In tetragonal complexes, the ground state is $d_{x^2-y^2}$. Hence, spin-orbit coupling is not possible and sharp EPR lines are observed.

EXERCISES

Exercise 4.1. In which region of the electromagnetic spectrum is EPR observed?

Exercise 4.2. Derivative curves are used in EPR instead of normal curves. Why? In the derivative curve, what is taken on the y-axis and what on the x-axis?

Exercise 4.3. The energy levels in EPR are reversed compared to those of NMR. Explain why.

Exercise 4.4. While in NMR positions of the signals are important, in EPR g factor is important. What could be the reason?

Exercise 4.5. What is the essential difference between *fine splitting* and *hyperfine splitting*?

Exercise 4.6. Hyperfine splitting in EPR provides valuable information. What is that?

Exercise 4.7. Why is $g_\perp$ more intense than $g_\parallel$?

Exercise 4.8. g is normally anisotropic. Explain.

Exercise 4.9. g can also be isotropic. When?

Exercise 4.10. Give the meaning of g_x and g_y. Under what circumstance does $g_x = g_y = g_\perp$?

Exercise 4.11. In what way is EPR more valuable than XRD with respect to paramagnetic complexes? (*Hint*: distortion)

Exercise 4.12. If an electron is present in a t_{2g} orbital, it will have orbital angular momentum, while an electron present in an e_g orbital will not. Explain.

Exercise 4.13. Crystal field splitting and spin-orbit coupling have opposite effects. Explain.

Exercise 4.14. What is the ground term in a Cu(II) complex? How is this split in a cubic field and tetragonal field? Explain by drawing the energy levels.

Exercise 4.15. In $[VO(H_2O)_6]SO_4$, how many d-electrons are there? How many lines will appear in its EPR spectrum?

Exercise 4.16. What is meant by zero-field splitting? How can you determine whether zero-field splitting has taken place or not?

Exercise 4.17. EPR spectra give concrete proof for metal-ligand coordination. How? Explain.

Exercise 4.18. Explain with suitable examples how the structure of a complex can be determined using EPR spectroscopy.

REFERENCES

[1] R.L. Dutta, A. Shyamal, Elements of Magnetochemistry, Affiliated East-West Press, New Delhi, 1993.

[2] B.N. Figgis, J. Lewis, The Magnetic Properties of Transition Metal Complexes, in: F. Cotton (Ed.), Progress in Inorganic Chemistry, vol. 6, Interscience, New York, 1967, pp. 37–239.

[3] R.S. Drago, Physical Methods for Chemists, second ed., Saunders (W.B.), Orlando, 1992.

CHAPTER 5

NMR Spectroscopy

Nuclear magnetic resonance abbreviated as NMR. The principles and applications of NMR are discussed in depth in various books [1–4]. NMR is shown by those nuclei whose nuclear spin, $I \neq 0$. The nuclear spin values for selected nuclei are given in Table 5.1.

5.1 PRINCIPLES OF NMR

It is well known that a spinning charged body produces a magnetic field. A nucleus is positively charged. Hence, if it has a spin too (ie, $I \neq 0$), it will produce a magnetic field. It behaves like a tiny bar magnet. This can be compared to a spinning top.

Nuclear spin, I, depends on the atomic and mass numbers of nuclei as shown below:

Mass No.	At. No.	Spin-Quantum No.
Odd	Odd or even	$1/2, 3/2, 5/2, \ldots$
Even	Even	0
Even	Odd	$1, 2, 3, \ldots$

5.1.1 Precessional Motion or Larmor Precession

A spinning top undergoes precessional motion in addition to its spin under the influence of the earth's gravitational field. Similarly, the spinning nucleus shows precessional motion in the presence of an external magnetic field. We have seen that the spinning top will be unstable and will fall down if it does not show precessional motion. In the same way, a spinning nucleus will be unstable if it does not undergo precessional motion. This is also known as Larmor Precession. This is shown in Figs. 5.1 and 5.2.

5.1.2 Precessional Frequency, ω

Precessional frequency (ω) is the frequency with which the nucleus rotates in an applied magnetic field, H. Precessional frequency is directly proportional to the applied magnetic field. That is, when the magnetic

Spectral Methods in Transition Metal Complexes. http://dx.doi.org/10.1016/B978-0-12-809591-1.00005-0

At. No.	Isotope	Abundance (%)	I	At. No.	Isotope	Abundance (%)	I
1	^{1}H	99.9850	1/2	23	^{51}V	97.75	7/2
1	^{2}H	0.0115	1	24	^{53}Cr	9.501	3/2
3	^{6}Li	7.59	1	25	^{55}Mn	100	5/2
3	^{7}Li	92.41	3/2	26	^{57}Fe	2.119	1/2
4	^{9}Be	100	3/2	27	^{59}Co	100	7/2
5	^{10}B	19.9	3	28	^{61}Ni	1.1399	3/2
5	^{11}B	80.1	3/2	29	^{63}Cu	69.17	3/2
6	^{13}C	1.07	1/2	29	^{65}Cu	30.83	3/2
7	^{14}N	99.632	1	30	^{67}Zn	4.10	5/2
8	^{17}O	0.038	5/2	46	^{105}Pd	22.33	5/2
9	^{19}F	100	1/2	47	^{107}Ag	51.839	1/2
11	^{23}Na	100	3/2	47	^{109}Ag	48.161	1/2
12	^{25}Mg	10	5/2	48	^{111}Cd	12.80	1/2
13	^{27}Al	100	5/2	48	^{113}Cd	12.22	1/2
14	^{29}Si	4.6832	1/2	50	^{115}Sn	0.34	1/2
15	^{31}P	100	1/2	50	^{117}Sn	7.68	1/2
17	^{35}Cl	75.78	3/2	50	^{119}Sn	8.59	1/2
17	^{37}Cl	24.22	3/2	78	^{195}Pt	33.832	1/2
21	^{45}Sc	100	7/2	79	^{197}Au	100	3/2
22	^{47}Ti	7.44	5/2	80	^{199}Hg	16.87	1/2
22	^{49}Ti	5.41	7/2	80	^{201}Hg	13.18	3/2
23	^{50}V	0.250	6	83	^{209}Bi	100	9/2

Table 5.1 Spin Value and Abundance of Selected Nuclei

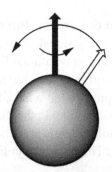

Figure 5.1 Precessional motion of a nucleus.

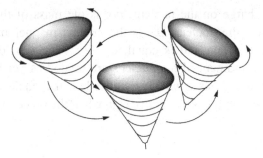

Figure 5.2 Spinning top.

Table 5.2 Magnetogyric Ratio of Selected Nuclei		
Nucleus	γ_N (10^6 rad $S^{-1} T^{-1}$)	$\gamma_N/2\pi$ (MHz T^{-1})
^{1}H	267.513	42.576
^{2}H	41.065	6.536
^{3}He	203.789	32.434
^{7}Li	103.962	16.546
^{13}C	67.262	10.705
^{14}N	19.331	3.077
^{15}N	−27.116	−4.316
^{17}O	36.264	5.772
^{19}F	251.662	40.052
^{23}Na	70.761	11.262
^{27}Al	69.763	11.103
^{31}P	108.291	17.235
^{57}Fe	8.681	1.382
^{63}Cu	71.118	11.319
^{67}Zn	16.767	2.669

field strength is increased, the precessional frequency also increases. The relation is

$$\omega = \gamma H \tag{5.1}$$

where ω is the precessional frequency, H is the applied magnetic field, and γ is called the *magnetogyric ratio*. This γ is a constant for a given nucleus. The value of γ for selected nuclei are given in Table 5.2. The value of γ is given by

$$\gamma = \frac{e}{2m_N} g_N = \frac{g_N \mu_N}{\hbar} \tag{5.2}$$

where e is the charge on the nucleus, m_N is the mass of the nucleus, g_N is the g-factor for the nucleus, μ_N is the nuclear magneton, and $\hbar$ is the reduced Planck constant and is equal to $h/2\pi$. The spin of the nucleus and the magnetic moment of the nucleus are related as per Eq. (5.3). The magnetogyric ratio γ has a *positive* sign when the nuclear precessional motion is *clockwise* and has a *negative* sign when the precessional motion is *counterclockwise*.

$$\mu = \gamma I \tag{5.3}$$

The reason behind the precessional motion of the nucleus is that the external magnetic field tries to bring the direction of the moment of the nucleus in line with its own direction. That is, the direction of the nuclear moment should align with the direction of the external magnetic field.

5.1.3 Energy Levels and Transition

When the spin of a nucleus is equal to I, and when $I \neq 0$, all the magnetic states of that nucleus will be degenerate; that is, they will have the same energy. However, when the nucleus is placed in an external magnetic field, only certain orientations are allowed. These allowed transitions are given by

$$m = I, I - 1, I - 2, \ldots, -I + 1, -I \tag{5.4}$$

This gives rise to $(2I + 1)$ orientations. These levels are split in the presence of a magnetic field. However, the transitions are allowed according to the selection rule, $\Delta m_I = \pm 1$. When a nucleus has $I = 1/2$, $2I + 1 = 2$. Thus there will be two energy levels for protons. The energy level with $m_I = +1/2$ results when the magnetic moment of the nucleus is aligned with the external magnetic field and will have lower energy. The other magnetic moment will be opposed to the external magnetic field and will have higher energy. These observations are shown in Figs. 5.3–5.5. Thus, there will be one transition and one signal in PMR.

Resonance

In the ground state, a proton has a spin of $1/2$ and in the excited state its spin changes to $-1/2$. Thus, during transition spin inversion takes place and it is called nuclear flipping. In order that the transition takes place, the proton must absorb energy, ΔE. The energy is supplied in the form of radio frequency v from an external source. If the proton undergoing precessional motion is to absorb this energy, then its precessional frequency ω must be equal to the radio frequency v.

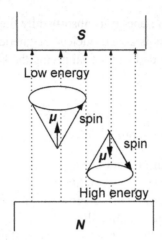

Figure 5.3 Nuclear spin and magnetic field.

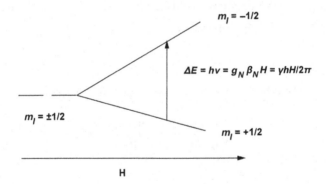

Figure 5.4 The energy levels in PMR.

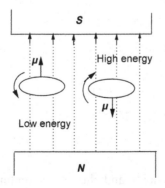

Figure 5.5 Current loop and magnetic moment.

In this condition, $\omega = \nu$, energy is absorbed by the proton and it goes to the excited state. This is called nuclear magnetic resonance and the absorbed energy is recorded as a signal giving the NMR spectrum.

Field Sweep and Frequency Sweep Methods
Resonance can be brought about in *two* ways:

1. The magnetic field can be kept constant, while the radio frequency can be changed until the resonance condition $\omega = \nu$ is met. This is called the *frequency sweep* method. This is difficult because the radio frequency source has to be changed continually.
2. In the other method, the radio frequency is kept constant while the magnetic field is continuously varied until the resonance condition is satisfied. This is called the *field sweep* method. This is easier because the strength of the magnetic field can be varied by varying the electric field (the magnet is an electromagnet).

Relation Between ν and ω
The relation between ω and ν is derived as shown below:

$$\omega = \frac{\text{Magnetic moment}}{\text{Angular momentum}} H \text{ rad s}^{-1} \tag{5.5}$$

$$= \frac{\mu H}{2\pi \mathbf{I}} \text{Hz} \tag{5.6}$$

$$\mu = \frac{g\beta_N \sqrt{I(I+1)}}{2\pi} \tag{5.7}$$

$$\mathbf{I} = \sqrt{I(I+1)} \frac{h}{2\pi} \tag{5.8}$$

$$\omega = \frac{g\beta_N \sqrt{I(I+1)} H/2\pi}{\sqrt{I(I+1)}(h/2\pi)} \tag{5.9}$$

$$= g\beta_N H/h \text{ Hz} \tag{5.10}$$

$$\nu = g\beta_N H/h \text{ Hz} \tag{5.11}$$

$$\therefore \omega = \nu \tag{5.12}$$

The precessional frequencies and the corresponding magnetic fields are given in Table 5.3.

Table 5.3 Precessional Frequencies of Selected Nuclei						
H (Tesla)	1.4	2.1	2.3	5.1	5.8	7.1
^1H	60	90	100	220	250	300
^2H	9.2	13.8	15.3	33.7	38.4	46.0
^{13}C	15.1	22.6	25.2	55.0	62.9	75.5
^{14}N	4.3	6.5	7.2	15.8	17.9	21.5
^{19}F	56.5	84.7	93.0	206.5	203.4	282.0
^{31}P	24.3	36.4	40.5	89.2	101.5	121.5

Each nucleus has a specific precessional frequency at a given magnetic field strength. Hence, while recording NMR spectra, signals of only one type of nucleus are seen. The signals of other nuclei are not seen. For example, if proton magnetic resonance (PMR) is recorded, only proton signals are seen and the signals of other nuclei such as fluorine are not seen even if they are present in the molecule. However, the splittings due to other nuclei such as fluorine are seen.

5.2 GROUND AND EXCITED STATE POPULATION

All the nuclei do not occupy the ground state because of thermal motion and Boltzmann distribution. Let us consider a nucleus of spin 1/2 giving rise to two energy levels. According to classical theory, at a temperature of T K, the ratio of populations of the two levels is given by

$$\frac{N_{upper}}{N_{lower}} = e^{-\frac{\Delta E}{kT}} \qquad (5.13)$$

where k is the Boltzmann constant. However, ΔE is extremely small. The ratio of populations can be calculated at a given temperature and at a given magnetic field as shown below:

$$\Delta E \approx 7 \times 10^{-26} \text{ J}$$

$$k = 1.38 \times 10^{-23} \text{ J K}^{-1}$$

$$T = 300 \text{ K}$$

$$\frac{N_{upper}}{N_{lower}} \approx e^{-\frac{7 \times 10^{-26}}{4.2 \times 10^{-21}}}$$

$$\approx e^{-1 \times 10^{-5}}$$

$$\approx 1 - (1 \times 10^{-5})$$

Thus the ratio is very nearly equal to unity. This means that the proton exists in the two spin states almost equally.

5.3 RELAXATION OF THE NUCLEI

Since the population difference in the two levels is very small, the population in the two states becomes equal quickly at the first excitation itself and immediately the signal should fade out. However, this does not happen in reality. We are able to get the signals continuously. How does this happen? This occurs by a process known as *relaxation* of the nucleus.

Relaxation is simply the return of the excited nucleus to the ground state.

Mechanism of Relaxation
The nucleus in the excited state loses its excess energy in *two* ways:

1. The excess energy is transferred to the surroundings (lattice). This is called *spin-lattice relaxation*.
2. The excess energy is transferred to the neighboring nucleus. This is called *spin-spin relaxation*.

5.4 SPIN-LATTICE RELAXATION (T_1)

Lattice refers to the surrounding atoms or molecules of solvent or the same molecule, etc. The nucleus in the excited state transfers its excess energy to the surrounding molecules which are vibrating and rotating. When they are properly oriented with respect to the excited nucleus, the transfer of energy takes place easily and quickly. This relaxation time is denoted as T_1. This is also known as *longitudinal relaxation* time.

5.5 SPIN-SPIN RELAXATION (T_2)

By this mechanism, the excess energy of the excited nucleus is transferred to the nucleus provided it has the same energy level as the excited one. In other words, spin-spin relaxation is a mechanism by which energy is mutually exchanged. When one nucleus loses energy, another nucleus gains energy. Hence, there is no net change in the populations of the two spinstates.

This relaxation time is denoted as T_2. This is also known as *transverse relaxation* time.

5.6 COMPARISON OF RELAXATION TIMES

The relaxation times vary in the case of solids and liquids.

T_1 relaxation is in the range 10^{-2}–10^4 s for solids and 10^{-4}–10 s for liquids. Liquids have shorter relaxation times because molecules move freely in liquids leading to larger fluctuations of magnetic fields in the vicinity of excited nuclei.

T_2 is very short for solids and is of the order of 10^{-4} s, and in the case of liquids $T_2 \approx T_1$.

5.7 WIDTH OF NMR LINES

When relaxation times, T_1 and T_2, are small, the excited nucleus reaches the ground state quickly and the signals will be broad. When relaxation times are large, the excited nucleus takes longer to reach the ground state and the signal will be sharp.

Explanation
Heisenberg's uncertainty principle: Let a nucleus be in an energy state for a definite time δt seconds. Now, the energy of that nucleus will be uncertain by an amount δE. Now,

$$\delta E \times \delta t \approx h/2\pi$$
$$E = h\nu$$
$$\delta E = h\delta\nu$$
$$h\delta\nu \times \delta t = h/2\pi$$
$$\delta\nu \times \delta t = 1/2\pi$$

Thus, $\delta\nu \times \delta t$ is a constant. From this relation it is clear that if δt is large, $\delta\nu$ must be small and vice versa. $\delta\nu$ is the uncertainty in frequency and is the signal width. If $\delta\nu$ is large the signal is broad and if it is small the signal is sharp. This is shown in Fig. 5.6.

Example: Signal of protons attached to nitrogen appear broad. Why? I for $N = 1$. When a nucleus has $I > 1$, it has a quadrupolar moment. This means

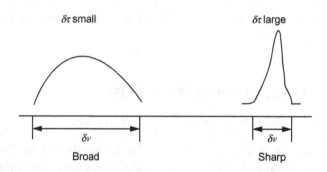

Figure 5.6 Signal width and relaxation time.

that the excited nucleus interacts both with the electric and magnetic field gradients and loses its excess energy quickly. Hence, the relaxation time is short and broad signals are obtained.

5.8 BASIC TYPES OF INFORMATION FROM NMR

Three basic types of information are obtained from NMR. They are given below:

1. chemical shift
2. splitting of signals and
3. integration of signals.

5.8.1 Chemical Shift

The separation between absorption peaks is called *chemical shift*. The same nuclei give rise to signals at different positions in different chemical environments. Hence, it is called chemical shift.

Origin of Chemical Shift
Applied and Effective Magnetic Field
The applied magnetic field is not completely perceived by the nucleus because the nucleus is covered by electrons. The electrons revolving around the nucleus produce a magnetic field, which opposes the applied magnetic field. This is called shielding. Hence, the magnetic field experienced by the nucleus will be less than that applied and the precessional frequency will also be less for these shielded protons. Therefore, the external magnetic field strength must be increased to increase the precessional frequency and cause resonance. Thus, the signal for a shielded proton will appear at

higher magnetic field strength. This shielding varies in different chemical environments. Electron-releasing groups increase the electron density at the nucleus and shielding will be accentuated. Hence, the effective field experienced by the nucleus will be less. Electron-withdrawing groups withdraw electrons from the nucleus and it will have greater exposure to the applied field. Hence, the effective field at the nucleus will be larger and the precessional frequency will be increased. Therefore, a smaller magnetic field is sufficient to induce resonance. This is called *deshielding*. It is shown in Figs. 5.7 and 5.8.

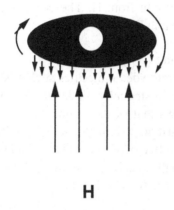

H

Figure 5.7 Shielding of a nucleus by surrounding electrons.

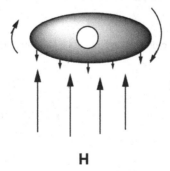

H

Figure 5.8 Deshielding of a nucleus by surrounding electrons.

Factors Affecting Chemical Shift

The chemical shift, that is, the position of signals, is influenced by *three* factors:

1. electronegativity
2. van der Waals deshielding and
3. anisotropic effects.

Electronegativity and Chemical Shift

Electronegativity of an atom is the capacity to attract the bonded pair of electrons toward itself. Thus Cl is more electronegative than C and hence withdraws electrons toward itself in a C–Cl bond. Hence, the electron density around C is reduced. To balance this, electron density in the C–H bond flows toward carbon from H. The net result is that the proton is deshielded. This is shown in Fig. 5.9.

van der Waals Deshielding

This is important in overcrowded molecules such as steroids, alkaloids, etc. When a proton occupies a sterically hindered position, the electron cloud of the bulky group repels that of the proton and the proton is deshielded and hence more exposed to the magnetic field. This is called van der Waals deshielding. This effect is significant only in overcrowded molecules because this effect is small. This is shown in Fig. 5.10. The hydrogen shown within the circle at position 10 is deshielded.

Anisotropic Effect and Chemical Shift

When a property does not depend on the direction, it is called isotropic. However, if a property varies depending upon the direction, the property is said to be *anisotropic*. This effect is due to the circulation of π electrons. When the π electrons rotate, a magnetic field is produced. This will be para-

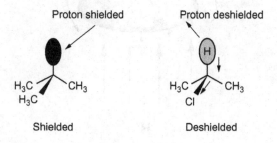

Figure 5.9 Deshielding of a proton due to electronegativity of chlorine.

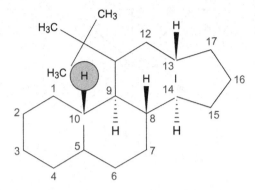

Figure 5.10 van der Waals deshielding.

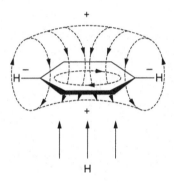

Figure 5.11 Anisotropic effect in benzene.

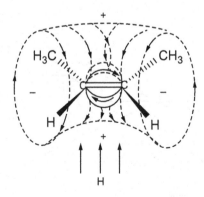

Figure 5.12 Anisotropic effect in ethylene.

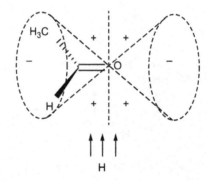

Figure 5.13 Anisotropic effect in an aldehyde.

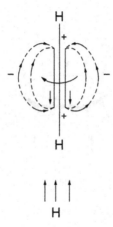

Figure 5.14 Anisotropic effect in acetylene.

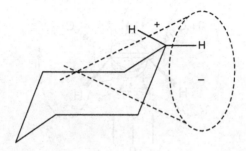

Figure 5.15 Anisotropic effect in cyclohexane.

magnetic, that is, aligned with the external magnetic field in one direction, and diamagnetic, that is, opposed to the external magnetic field in another direction. Hence, it is called anisotropic and is shown in Figs. 5.11–5.15. In all these figures, a plus sign indicates a shielded region and a minus (−) sign indicates a deshielded region.

Measurement of Chemical Shift

δ or τ Scale

Most commonly the chemical shift, that is, the positions of signals, are measured on the δ scale and less commonly on the τ scale. The two scales are related as

$$\tau = 10 - \delta$$

The δ scale is defined as

$$\delta = \frac{\text{Chemical shift in Hz}}{\text{Operating frequency of the instrument in MHz}}\ \text{ppm} \qquad (5.14)$$

Thus, the δ scale is dimensionless and it is merely a number.

Example 1: Let us record the PMR spectrum of a sample at 300 MHz with TMS as the internal standard. Let the external magnetic field strength be 70,500 gauss. Let the signal of the proton under consideration appear at 1100 Hz. Then what is the chemical shift? Substituting the values in Eq. (5.14), we get

$$\delta = \frac{1100}{300}$$
$$= 3.67\ \text{ppm}$$

The signal of the proton under consideration appears at 1100 Hz at a magnetic field strength of 70,500 gauss at radio frequency of 300 MHz. Under these conditions if the magnetic field strength is increased, the signal position will also change. In short, the signal position is dependent upon the magnetic field strength. However, when the signal position in hertz is divided by the operating frequency of the instrument, the value will be independent of the external magnetic field strength. Hence, the δ value will be independent of the applied magnetic field strength.

Shielding, Deshielding, and δ-Value

The nomenclature used to describe the signal positions is detailed in Fig. 5.16 and the PMR spectrum of ethanol is given in Fig. 5.17.

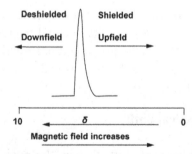

Figure 5.16 The terms used in the case of a PMR signal.

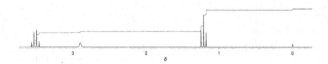

Figure 5.17 PMR spectrum of ethanol.

Internal Standard in NMR

Tetramethylsilane (TMS) is taken as the internal standard. The reasons are given below:

1. It has 12 hydrogens and hence even a small amount of TMS gives a relatively large signal.
2. Since all the hydrogens are equivalent, there will be singlets only and hence the signal of TMS can be easily identified.
3. The structure of tetramethylsilane is shown in Fig. 5.18. Si is less electronegative than carbon and hence pushes electron density toward carbon, which in turn increases the electron density around the hydrogens. Thus, the hydrogens are very much shielded and the signal appears upfield. In this region, signals of protons from other compounds do not interfere.
4. It is inert like an alkane and hence will not react with compounds dissolved in it.
5. It is volatile (b.p. 27°C). Hence, TMS can be easily evaporated to get back the original compound.

5.8.2 Splitting of Signals

This is very important information obtained from NMR regarding the structure of a compound. The splitting pattern gives all the necessary

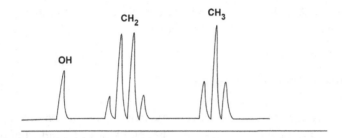

Figure 5.18 Structure of TMS.

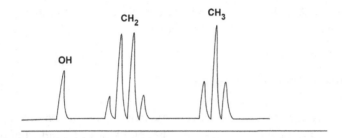

Figure 5.19 Splitting pattern of ethyl group.

information about the structure. For example, a triplet-quartet splitting pattern shows the presence of an ethyl group as shown in Fig. 5.19.

Theory of Splitting

Why should a signal be split in NMR? It is due to the interaction with the spin of the neighboring nuclei. When a nucleus has spin I, it will have $2I + 1$ orientations. Hence, a proton H_A will have $2 \times 1/2 + 1 = 2$ orientations if there is one neighboring proton. If there are two neighboring protons, then proton H_A will have $2 \times 2 \times 1/2 + 1 = 3$ orientations. In general, if there are n equivalent protons, then the signal of H_A will have $2nI + 1$ orientations. The proton H_A will resonate for each orientation and will give a signal for each resonance.

$(n + 1)$ Rule

Thus in the $-CH_2CH_3$ group, the CH_2 protons neighbor *three* methyl protons and hence have $2 \times 3 + 1/2 + 1 = 4$ orientations and the signal of methylene protons is split into four, called a *quartet*. In the same manner, the methyl protons neighbor *two* methylene protons and hence resonate $2 \times 2 \times 1/2 + 1 = 3$ times and the methyl signal is split into a *triplet*. In general, when there are n neighboring protons, the signal of an adjacent proton will be split into

$2 \times n \times 1/2 + 1 = (n + 1)$ signals. Hence, the splitting pattern in PMR can be easily deduced using the $(n + 1)$ rule.

Coupling Between Nuclei and Coupling Constant, J

The magnetic field of one proton interacts with that of another. This is called spin-spin coupling. When this coupling is between chemically nonequivalent protons, it leads to splitting of signals into doublet, triplet, quartet, etc., and the separation between the multiplets is called the coupling constant, J.

Separation between signals is called chemical shift, δ, and the separation between multiplets is called the coupling constant, J.

Chemical and Magnetic Equivalence

Chemical Equivalence of Protons

If two protons have the same connectivity, that is, they are connected to the same atoms or groups by the same number and kinds of bonds, the protons are said to be *chemically equivalent*. This is shown in Fig. 5.20. Chemically nonequivalent protons (H_a and H_b) are shown in Fig. 5.21. The relation between chemical symmetry and chemical equivalence is shown for toluene in Fig. 5.22.

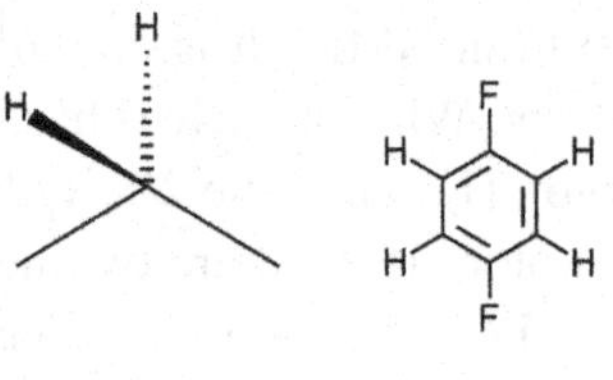

Figure 5.20 Chemically equivalent protons.

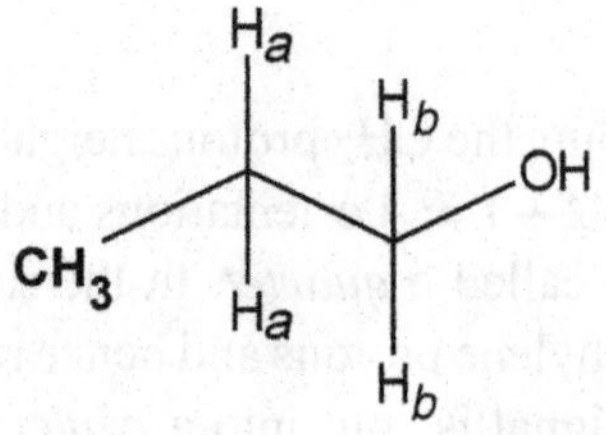

Figure 5.21 Chemically nonequivalent protons.

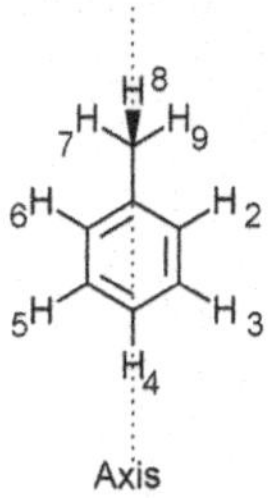

Rotation by 180° interconverts 6 & 2 and 5 & 3, hence equivalent.

Rotation by 120° shifts 8 to 9, 9 to 7, and 7 to 8 and hence equivalent.

Rotation by 240° shifts 7 to 9, 8 to 7, and 9 to 8, hence 7,8,and 9 are equivalent.

Axis

Figure 5.22 Chemical equivalence and symmetry.

Magnetically Equivalent and Nonequivalent Protons

If both the chemical environment and coupling constants (J) are the same, the nuclei are said to be both chemically and magnetically equivalent. If the chemical environment is the same but the J values are different, then the nuclei are chemically equivalent but magnetically nonequivalent.

Thus, magnetically equivalent nuclei are also chemically equivalent but the reverse is not necessarily true.

How to Identify Chemically Equivalent Protons?

1. Nuclei exist in *identical* environments.
2. A pair of chemically equivalent nuclei can be interchanged by any symmetry operation or by rapid interchange compared to the NMR time scale.
3. A pair of protons are called homotopic protons if they can be interchanged by rotation about an axis, C_n. The protons in dichloromethane can be cited as an example of this.
4. If a pair of protons cannot be interchanged by a proper axis of rotation but only by an improper axis of rotation (S_n), they are called enantiotopic protons. For example, the protons attached to C_α of glycine are enantiotopic.
5. If the *geminal* protons (CH_2) cannot be interchanged by symmetry operation, they are called *diastereotopic* protons and these protons are *not* chemically equivalent. The β-methylene protons, where $-CH_2$ protons are attached to C_α, atoms belong to this category. These are shown in Figs. 5.23–5.25.
6. When a structure undergoes rapid interconversion by rotation about bonds, for example, one chair conformation to another chair conformation or due to chemical changes such as keto-enol tautomerism, the protons will be chemically equivalent because the interconversion is fast compared to the NMR time scale and hence only one averaged orientation is seen by the instrument. This is shown in Fig. 5.26.

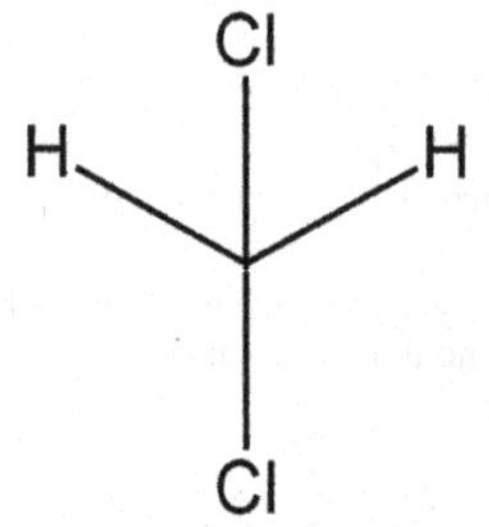

Figure 5.23 Homotopic protons.

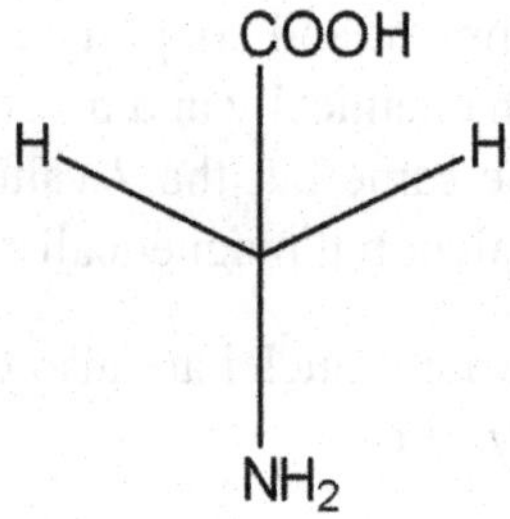

Figure 5.24 Enantiotopic protons.

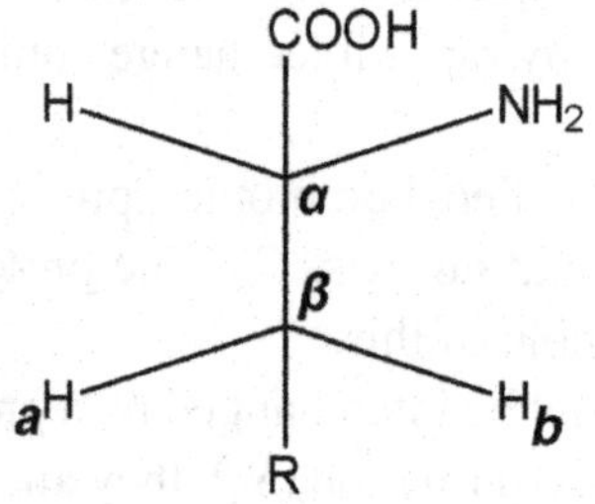

Figure 5.25 Diastereotopic protons.

Figure 5.26 Keto-enol tautomerism.

How to Identify Magnetically Equivalent Protons?
Magnetic equivalence of nuclei can be determined as follows:

1. First, the nuclei under consideration should be chemically equivalent.
2. There should be spin-coupling equivalence.
3. Each member of the group of nuclei should have the same coupling interaction (J should be equal) to every other nucleus in the molecule. Examples of magnetically nonequivalent protons are shown in Fig. 5.27. In Fig. 5.27B, $J_{ab} \neq J_{ab'} \neq J_{aa'}$. Hence, the four protons are *not* magnetically equivalent.

> If nuclei are to be tested for magnetic equivalence, they must first be chemically, that is, chemical shift equivalent. In other words, if nuclei are *not* chemical shift equivalent, they will never be magnetically equivalent. Chemical shift equivalent nuclei will be magnetically equivalent if they couple to other nuclei in the set equally. In short, magnetically equivalent nuclei will be chemical shift equivalent. However, the reverse need not be true. That is, chemical shift equivalent nuclei need not be magnetically equivalent.

Other examples of magnetic nonequivalence are shown in Fig. 5.28.

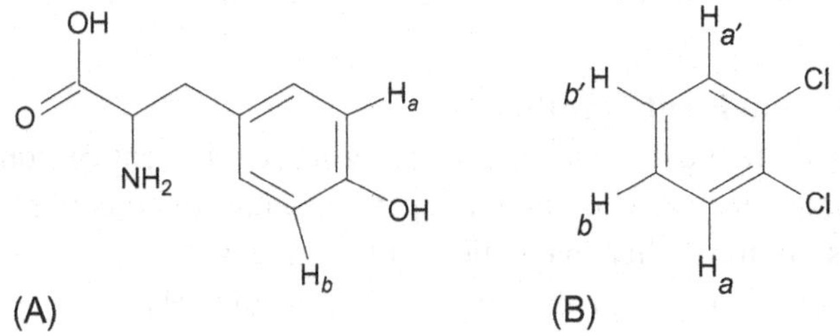

Figure 5.27 Magnetically nonequivalent protons.

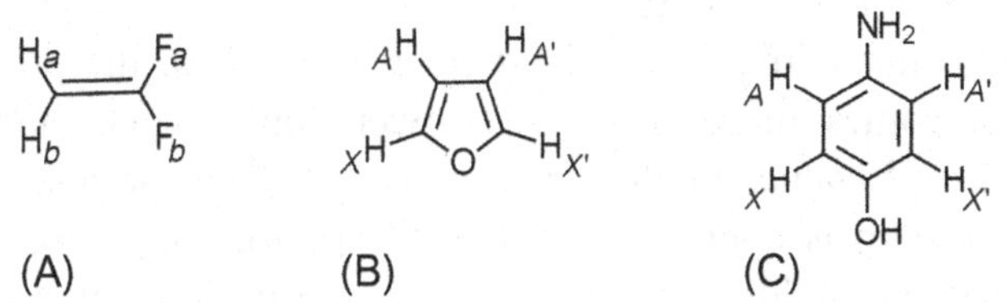

Figure 5.28 Magnetically nonequivalent nuclei.

In Fig. 5.28A, the two H_s atoms are chemically equivalent and the two F_s atoms are chemically equivalent because $\delta_{H_a} = \delta_{H_b}$ and $\delta_{F_a} = \delta_{F_b}$. However, from the geometry of the molecule, $J_{H_aF_a} \neq J_{H_aF_b}$ and $J_{H_bF_a} \neq J_{H_bF_b}$. In Fig. 5.28B, $J_{XA} \neq J_{XA'}$ and $J_{H_{X'A'}} \neq J_{H_{X'A}}$. Hence, the four H_s atoms are not magnetically equivalent. Similarly, in Fig. 5.28C, $J_{H_XH_A} \neq J_{H_XH_{A'}}$ and $J_{H_{X'}H_{A'}} \neq J_{H_{X'}H_A}$. Hence, the four hydrogens are not magnetically equivalent.

A simple method of testing for magnetic equivalence is based on the geometric relationship. That is, bond distances and bond angles from each nucleus in relation to the probe nucleus must be identical. In other words, if the two nuclei under consideration can be interchanged through a reflection plane passing through the probe nucleus and perpendicular to a line joining the chemical shift equivalent nuclei, they will be magnetically equivalent.

This is explained in Fig. 5.29A-C.

There is a subtle difference between spin-spin coupling and spin-spin splitting. Spin-spin coupling occurs between sets of protons that have the same chemical shift. However, this coupling leads to splitting of signals only when the protons have different chemical shifts.

Splitting of Signals Into Multiplets

Consider two chemically and magnetically nonequivalent protons H_A and H_X. The resonance of H_A, for example, depends on the total magnetic field at this proton. This magnetic field is due to the effective field at the proton and that due to the neighboring proton, H_X. The contribution of the field from the neighboring proton depends on its orientation. The proton H_X has *two* orientations, namely aligned with and opposed to the field of H_A. When the two fields are aligned, the magnetic field of the neighboring proton contributes and the effective field at H_A increases and the precessional frequency will be increased. Now, the proton under consideration comes to resonate at a lower applied field and gives a signal at lower δ value. In the other orientation, the orientation of the neighboring proton is opposed to that of the proton under consideration. Now, the magnetic fields of the two protons are in opposition and hence the effective magnetic field at the proton under consideration, H_A, is less.

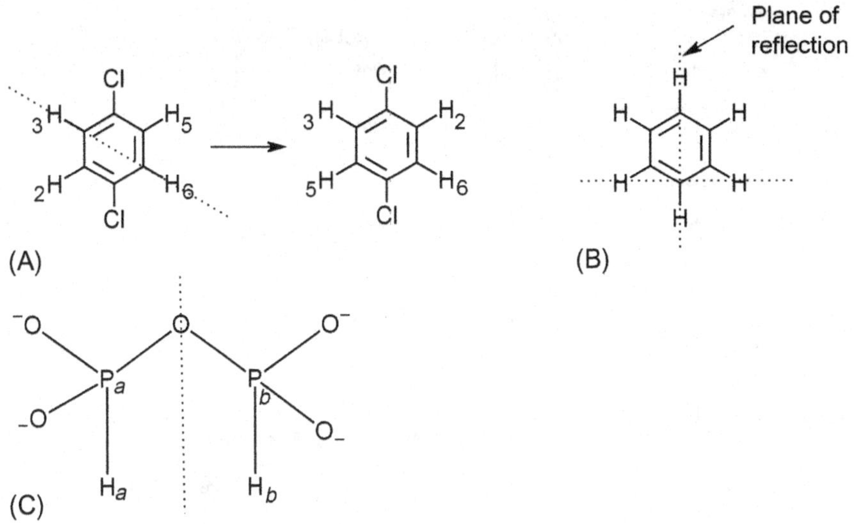

Figure 5.29 Test for magnetic equivalence by geometric relationship. (A) H_2 and H_6 are chemically equivalent but are not interchanged by reflection. Hence, they are not magnetically equivalent. $J_{2,3} \neq J_{2,5}$ and $J_{6,5} \neq J_{6,3}$. Hence, H_2 and H_6 are not magnetically equivalent. (B) Protons are magnetically equivalent. (C) P_a and P_b are chemically equivalent because they can be interconverted by 180° rotation about the axis shown passing through the center oxygen or reflection about this plane. For the same reason, H_a and H_b are chemically equivalent. But P_a and P_b are not magnetically equivalent because the geometric relation of the two is not the same. P_a and P_b are to be tested for magnetic equivalence. Join these two by a line. Probe nucleus is H_a. Imagine a plane containing H_a and perpendicular to the line joining P_a and P_b. Check if P_a and P_b can be interchanged by reflecting in this plane. They cannot be interchanged. Hence, P_a and P_b are not magnetically equivalent. In the same way, H_a and H_b are not magnetically equivalent.

Therefore, the precessional frequency will be decreased. Hence, the proton will exhibit resonance at a higher applied magnetic field. Thus a signal will be obtained at a slightly higher magnetic field. In other words, the proton H_A resonates twice, giving rise to two signals. That is, the signal of H_A appears as a doublet due to the coupling with one neighboring proton. In this way the spitting due to n neighboring protons can be explained.

Intensity of Multiplets

When there is one adjacent proton, it can have two orientations corresponding to $m_I = +1/2$ and $-1/2$. Both orientations have equal probability. Hence, the intensity will be 1:1. When there are two protons, $I = +1$. Then the possible orientations will correspond to $m_I = +1, 0,$ and -1. These will be in the ratio 1:2:1 as explained in Fig. 5.30.

Pascal's Triangle The intensities of multiplets are obtained easily using Pascal's triangle as shown in Fig. 5.31.

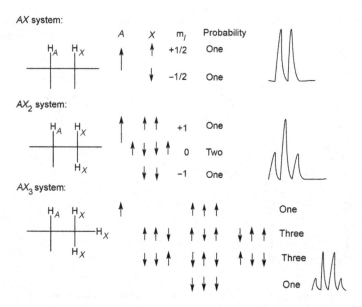

Figure 5.30 Intensity of multiplets.

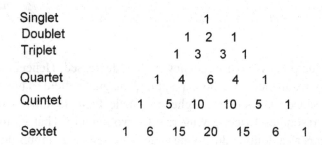

Figure 5.31 Pascal's triangle and intensity of multiplets.

Chemical Shift, δ and Coupling Constant J

The differences between δ and J are shown in Fig. 5.32.

The differences between δ and J are summarized below:

1. While δ is the separation between signals, J is the separation between multiplets.
2. While δ depends upon the external magnetic field strength, J is independent of the external magnetic field strength.

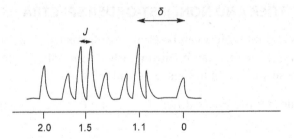

Figure 5.32 Difference between δ and J.

5.8.3 Integration of the Signal

Another important piece of information obtained from NMR is the integration of the signals. This gives the area under each signal. Since the area under a signal is proportional to the number of nuclei giving rise to the signal, we can find the number of nuclei belonging to each signal from the integration of the spectrum. It is simply the stepwise curve above each signal and is shown in Fig. 5.33.

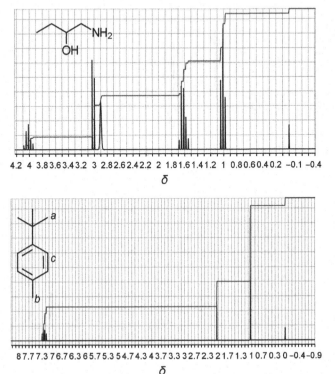

Figure 5.33 Integration of NMR spectrum.

5.9 FIRST-ORDER AND NONFIRST-ORDER SPECTRA

If the signals and multiplets can be easily distinguished, it is called a first-order spectrum. If neither can be distinguished, it is called a nonfirst-order spectrum. These are shown in Figs. 5.34 and 5.35.

The δ-values for protons in different environments are given in Table 5.4.

5.10 J-VALUE FOR DIFFERENT GEOMETRIES

The separation between the multiplets is known as the coupling constant, J. It is expressed in hertz. For example, if a doublet is formed due to the interaction of protons A and X, then the coupling is represented as J_{AX}. The

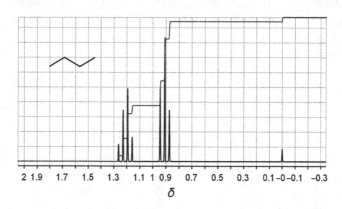

Figure 5.34 First-order NMR spectrum.

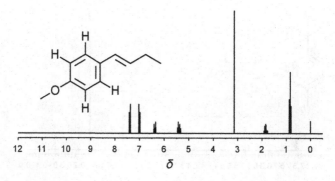

Figure 5.35 Nonfirst-order spectrum.

Table 5.4 δ-Values for Protons in Different Environments			
Functional Group	**δ (ppm)**	**Functional group**	**δ (ppm)**
RCH_3	0.9	Ether, H–C–OR	3.3–4.0
R_2CH_2	1.3	Esters, RCOO–C–H	3.7–4.1
R_3CH	1.5	RCOOH	10.5–12.0
C=C–H	4.6–5.9	RCHO	9–10
C≡C–H	2.3	H–C–C=O	2.0–2.7
Ar–H	6.0–8.5	Ar–O–H	4–12
Benzylic, ArC–H	2.2–3.0	C=C–O–H	15–17
Allylic, C=C–C–H	1.7	RNH_2	1–5
Alcohol, ROH	3.4–4.0		

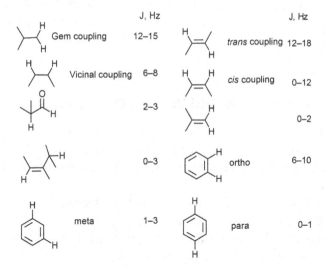

Figure 5.36 J-values for protons in different environments.

magnitude depends the internal arrangement of various protons and not on the external magnetic field. A few selected J-values are given in Fig. 5.36.

5.11 FACTORS INFLUENCING GEMINAL COUPLING, J_{gem}

There are two main factors that affect J_{gem}:

1. Electronegativity of the attached substituent.
2. H–Ĉ–H bond angle.

Electronegativity

The electronegativity of the attached substituent alters the value of J_{gem}. In $-CH_2X-$, the gem coupling varies with the nature of the substituent (electronegativity), X, from 12 to 9 Hz. The couplings cannot be measured directly because both the protons have the same δ-value. However, when one H is replaced by D, we get $-CHD-X$ and the gem coupling between H and D can be measured. $J_{H,H}$ can now be calculated as:

$$J_{H,H} = 6.53 J_{H,D} \tag{5.15}$$

Bond Angle

When the H–$\hat{C}$–H bond angle varies, J_{gem} varies. When the angle strain increases, J_{gem} decreases. For example, cyclohexanes and cyclopentanes are strain free and $J_{gem} = 10\text{–}14$ Hz. Cyclobutanes have $J = 8\text{–}14$ Hz and cyclopropanes have $J = 4\text{–}9$ Hz. Thus, when the angle strain increases, J_{gem} decreases.

5.12 FACTORS INFLUENCING VICINAL COUPLING, J_{vic}

1. Electronegativity of the attached substituent alters J_{vic} as in the case of J_{gem}. The greater the electronegativity, the smaller the J. For example, unhindered ethanes have $J \approx 8$ Hz, while the haloethanes have $J \approx 6 - 7$ Hz.
2. If there is restricted rotation, then the angle subtended by the electronegative substituent at the C–C bond influences J_{vic}.
3. Factors which affect the bond angles H–$\hat{C}$–C and C–$\hat{C}$–H will influence J_{vic} and this will be significant in smaller rings.

Karplus Equation

The dihedral angle (ϕ) is shown in Fig. 5.37.

4. The Karplus equation is given in Eq. (5.16) for ϕ between 0° and 90°:

$$J_{vic} = 8.5 \cos^2 \phi - 0.28 \tag{5.16}$$

If ϕ is between 90° and 180°, the Karplus equation is given by

$$J_{vic} = 9.5 \cos^2 \phi - 0.28 \tag{5.17}$$

According to these equations, J_{vic} is maximum when $\phi = 180°$ or 0°.

5. The Karplus equation is represented pictorially as in Fig. 5.38.

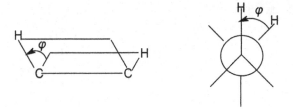

Figure 5.37 Dihedral angle φ.

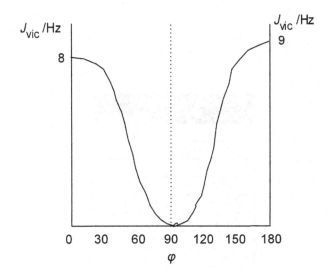

Figure 5.38 Pictorial representation of Karplus equation.

5.13 J-VALUE FROM THE SPECTRUM

In order to determine the J-value from the NMR spectrum, the separation between a multiplet of a signal in ppm is determined and this is then multiplied by the operating frequency of the instrument in MHz. Now, the J-value is obtained in Hz. Consider the spectrum shown in Fig. 5.39. Let it be obtained at 200 MHz. The J-value is determined as:

$$J = (3.510 - 3.475) \text{ ppm} \times 200 \text{ MHz} = 7 \text{ Hz} \qquad (5.18)$$

Some typical J-values are given in Table 5.5 [3].

The superscript in Table 5.5 gives the number of bonds between the coupling nuclei. Thus ^1J means that there is one bond between the two

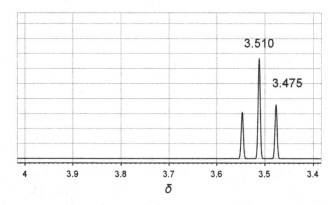

Figure 5.39 J-value from spectrum.

Table 5.5 Some Typical J-Values	
J	**Magnitude (Hz)**
$^2J(^1H,^{13}C)$ in C_2H_6	125
$^2J(^1H,^{13}C)$ in C_2H_4	160
$^2J(^1H,^{13}C)$ in C_2H_2	250
$^1J(^{19}F,^{31}P)$	500–1500
$^2J(^{19}F,^1H)$	10–50
$^1J(^{11}B,^1H)$ terminal	100–200
$^1J(^{11}B,^1H)$ bridging	<100

Table 5.6 J-Values, Oxidation State, and Hybridization	
J	**Magnitude (Hz)**
$^1J(^{31}p^V,^1H)$	400–1100
$^1J(^{31}p^{III},^1H)$	100–250
$^2J(^1H,^1H)$ sp^3-C	10–20
$^2J(^1H,^1H)$ sp^2-C	2–10

nuclei, ^2J means that there are two bonds, ^3J means that there are three bonds between the nuclei and so on. J-values depend upon the oxidation state and also the nature of hybridization. This is shown by the values given in Table 5.6.

5.14 RELATION BETWEEN J AND γ

$$\nu = \frac{\gamma H}{2\pi} \tag{5.19}$$

$$\gamma = \frac{2\pi\mu}{hI} \tag{5.20}$$

$$\omega_0 = \gamma H \tag{5.21}$$

$$\gamma H = 2\pi\nu \tag{5.22}$$

$$\omega_0 = 2\pi\nu \tag{5.23}$$

$$\frac{J_{AB}}{J_{CB}} = \frac{\gamma_A}{\gamma_C} \tag{5.24}$$

$$\frac{J_{HX}}{J_{DX}} = \frac{\gamma_H}{\gamma_D} \tag{5.25}$$

5.15 LABELING THE SPIN SYSTEMS

Systems can be labeled as AX, AB, AMX, or ABC based on the number of couplings and the distance between the signals similar to the distance between the letters A, B, C, M, X, etc. When there are only two signals and if they are well separated, it is known as an AX spectrum. If the two signals are very close, it is called an AB spectrum. If there are three signals well separated it is called an AMX spectrum and if they lie close it is called an ABC system. Different spin labeling systems are explained as follows:

AX System
When chemical shifts are very much greater than the coupling constants, an AX system results. In other words, $\nu_{AX} \gg J_{AX}$.

AB System
When $\nu_{AB} < 5J_{AX}$, the system is called an AB system.

AMX, ABX, and ABC Systems
An AMX system shown in Fig. 5.40.

The splitting pattern is explained as shown in Fig. 5.41. There are three coupling constants. Proton-5 is coupled with proton-4 ($J_{54} = 1$ Hz) and with proton-3 ($J_{53} = 1$ Hz. Hence, proton-5 shows two pairs of peaks. Proton-3 couples with proton-4 ($J_{34} = 3.5$ Hz) and with proton-5 ($J_{35} = 1$ Hz). Hence, proton-3 shows a doublet of doublets. Proton-4 is coupled with proton-3

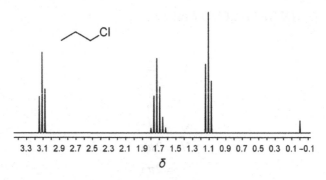

Figure 5.40 AMX system.

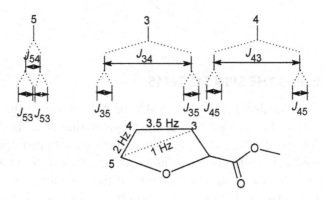

Figure 5.41 Splitting in an AMX system.

(J_{43} = 3.5 Hz) and with proton-5 (J_{45} = 2 Hz). This gives rise to a doublet of doublets.

Examples and splittings for ABX or ABC systems are shown in Figs. 5.42–5.45.

5.16 APPLICATIONS OF NMR

5.16.1 Inorganic Compounds

NMR spectra of some selected inorganic compounds are given in Fig. 5.46.

NMR spectra of inorganic compounds can also be predicted and sketched. A few examples are shown in Fig. 5.47. One point to be kept in

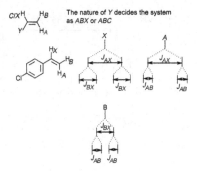

Figure 5.42 Examples and splittings for ABC and ABX systems.

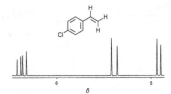

Figure 5.43 ABX NMR spectrum.

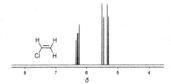

Figure 5.44 ABC NMR spectrum.

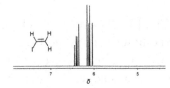

Figure 5.45 ABC NMR spectrum.

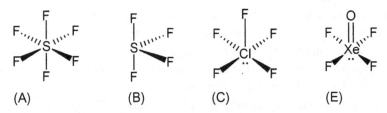

Figure 5.46 NMR signals of selected inorganic compounds. (A) One signal in ^{19}F NMR. (B) Two signals in ^{19}F NMR. (C) Two signals in ^{19}F NMR. (D) One signal in ^{19}F NMR.

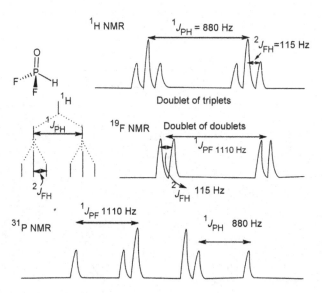

Figure 5.47 Splitting pattern in HPOF$_2$.

mind is that the coupling constant, J, decreases significantly with distance (ie, the number of bonds).

Example 1: The structure of HPOF$_2$ and the predicted NMR are shown in Fig. 5.47.

Example 2: The NMR spectrum of [H$_2$P$_2$O$_5$]$^{2-}$ is sketched in Fig. 5.48 [3].

Example 3: cis-[Pt(PEt$_3$Cl$_2$] [3]. Its structure and spectra are discussed in Fig. 5.49. The important point to be noted in this example is that platinum has five stable isotopes with mass numbers 192, 194, 195, 196, and 198. Of these five isotopes, only one isotope, ^{195}Pt, has $I = 1/2$ (abundance = 33.83%), while others have $I = 0$. When the ^{31}P NMR spectrum of the compound is recorded, the signal of P is split by this NMR-active nucleus into a doublet to the extent of the percentage abundance of the active nucleus present in this compound, while the inactive nucleus is not splitting the P signal and a singlet will be obtained. The singlet will be more intense than the doublet in proportion to the percentage abundance of the active and inactive nuclei and the signals will be superimposed as shown in Fig. 5.49. The intensity of the singlet will be approximately 64% and that of the doublets combined will be 34%. Hence, the intensity of each signal of the doublet will be equal to $34/2 = 17\%$. In order to make the spectrum simple,

$[H_2P_2O_5]^{2-}$

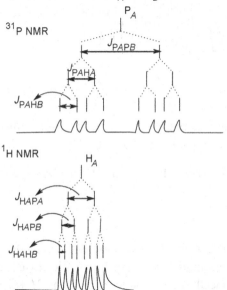

P_A and P_B are chemically equivalent but not magnetically beacuse they couple differently with other chemically equivalent nuclei, viz., the protons. Hence, signal of P_A is split by P_B and vice versa. Similarly, H_A and H_B are chemically equivalent but magnetically non-equivalent because they couple differently with the chemically equivalent nuclei, viz, P_A and P_B. Hence, the signal of P_A is split by $P_{B'}$ H_A and H_B and that of P_B is split by $P_{A'}$ H_A and H_B as shown here.

Figure 5.48 The splitting pattern in $[H_2P_2O_5]^{2-}$.

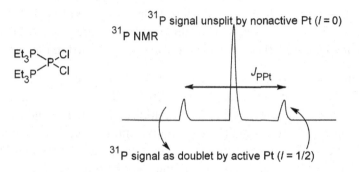

Figure 5.49 cis-[Pt(PEt_3)_2Cl_2] spectrum.

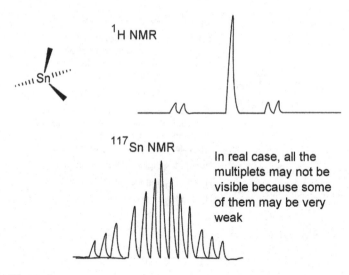

Figure 5.50 ^{117}*Sn and* ^{1}*H NMR spectra of tetramethyltin.*

the coupling due to protons is removed and the spectrum shown is a proton decoupled spectrum. This is represented as $^{31}P\{^1H\}$.

Example 4: Tetramethyltin. The ^{1}H NMR and ^{117}Sn NMR spectra are shown in Fig. 5.50 [3].

$I>1/2$ and NQR

When $I > 1/2$, a nuclear quadrupole moment results leading to line broadening due to the additional transitions possible. In general, when a nucleus with spin I is placed in a magnetic field, it has $2I + 1$ orientations, namely the m_I values. The m_I values are $I, I - 1, \ldots, -(I - 1), -I$. ^{10}B has $I = 3/2$. The possible m_I values are $3/2, 1/2, -1/2, -3/2$. As per the selection rule, $\Delta m_I = \pm 1$, the allowed transitions will be $3/2 \leftrightarrow 1/2$, $1/2 \leftrightarrow -1/2, -1/2 \leftrightarrow -3/2$. Thus three transitions are possible. This can be explained taking BH_4^- as an example that is explained in Fig. 5.51 [3].

Boron has two isotopes, ^{10}B and ^{11}B. For both the isotopes I is $\neq 0$. ^{10}B has $I = 3$ with an abundance of 20%. ^{11}B has $I = 3/2$ with an abundance of 80%. BH_4^- anion has a tetrahedral structure and hence all the four protons attached to boron are equivalent. Hence, there will be one signal in the ^{1}H NMR spectrum. However, this signal will be split by ^{11}B into $2 \times 1 \times 3/2 = 4$ lines. All these lines will be of equal intensity since these different orientations have equal probability. Similarly, this proton will

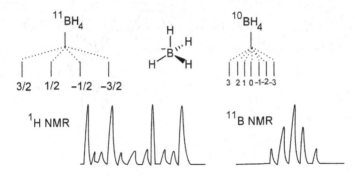

Figure 5.51 Splitting pattern and ^1H NMR of BH_4^- ion.

couple with ^{10}B and will be split into $2 \times 1 \times 3 + 1 = 7$ lines. Again all these lines will have equal intensity because all the orientations are equally probable. Since, ^{10}B has 20% abundance, the intensity will be divided amongst the seven lines and hence each line will have about $20/7 \approx 3\%$ intensity. If we want to record boron NMR we can have either ^{10}B NMR or ^{11}B NMR at a given time because the γ values are different (but splitting can be observed). We will get one line due to boron resonance because both the isotopes are in the same environment. However, this couples with four protons giving a quintet with the intensity ratio 1:4:6:4:1 as shown in Fig. 5.51. Since the abundance of this isotope is 80%, it will be shared by the five lines. Similarly, if we record the ^{10}B NMR spectrum, one line will be obtained which will be split into a quintet due to four protons. Since the intensity of this isotope is equal to 20%, this intensity will be shared by the five lines.

5.16.2 Metal Complexes

NMR can be applied to confirm complex formation and to find unequivocally which atom has coordinated to the metal from the change in δ-values. For example, if the ligand is a Schiff base, then its PMR spectrum is recorded. After complexation with a metal such as zinc [5], the PMR spectrum of the complex is once again recorded. The two spectra are compared. If there is any downfield shift in the δ-value of the proton attached to nitrogen, then we can confirm that the nitrogens of the Schiff base have coordinated to the metal as shown in Fig. 5.52. ^{113}Cd ($I = 1/2$) NMR is used in the study of metalloproteins [6]. The δ-value is affected by the nature of the ligands, the number of ligand, coordinating atom, bond length, coordination geometry, etc.

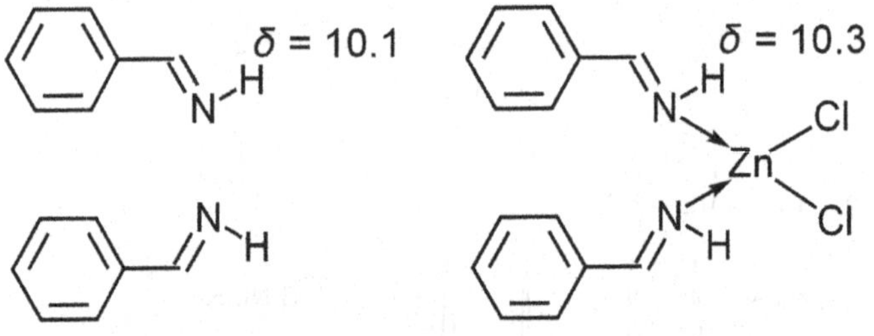

δ = 10.1   δ = 10.3

Figure 5.52 NMR and Schiff base Zn(II) complex.

The coordination chemistry of different gold complexes used as catalysts has also been studied by NMR [7]. Kintzinger et al. [8] have studied the ^{35}Cl NMR of chloride anion cryptates.

5.17 SOLID-STATE NMR

In this technique, NMR spectra of solid samples can be taken. Hence, this technique will be very useful when the sample is not soluble in solvents. This may be the case for polymers, macromolecules, inorganic compounds, complexes, etc. [9]. In solution spectra, the signals will be sharp because of rapid movement of molecules resulting in an averaged orientation. However, in the solid state, the positions are frozen and there is no averaged orientation. Hence, they will be anisotropic (vary with directions) and the effect will be signal broadening. Thus, the signals obtained will be very broad in solid-state NMR as shown in Fig. 5.53. Hence, in order to reduce the broadening and get useful information from the solid-state spectrum, some special techniques such as magic angle spinning (MAS), cross-polarization (CP), etc., are used.

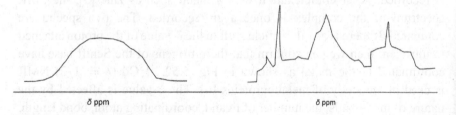

δ ppm   δ ppm

Figure 5.53 Static solid-state NMR.

Magic Angle Spinning

Dipolar coupling and chemical shielding depend on the term $3\cos^2\theta - 1$. Hence, this term should vanish if we want to remove these effects causing line broadening. $3\cos^2\theta - 1$ becomes equal to zero when $\theta = 54.74°$. This angle is called the *magic angle*. That is, when the sample is placed at this angle with respect to the external magnetic field, the line broadening can be reduced.

Spinning the Sample
In addition to this magic angle, the sample has to be spun so that average orientation results and the line broadening is eliminated giving sharp signals. The spin rate should be greater than the dipolar line width, which may be many kilohertz wide. When the magic angle and spinning are combined, it is called MAS. If the rate of spinning is less, additional side lines appear and complicate the spectrum. Thus the positions and number of side bands depend upon the spinning frequency. However, the position of the isotropic center band position remains the same whatever the spinning frequency and hence can be easily identified.

The rate of MAS should be $\geq$ magnitude of anisotropic interaction to get sharp signals.

Technique
The sample is finely powdered and tightly packed into a rotor and the rotor is spun at rates from 1 to 35 kHz depending on the rotor size and type of experiment to be conducted. The technique and the effect of spin rate on the spectrum are shown in Fig. 5.54.

Cross-Polarization

Here, polarization from spins of abundant nuclei such as ^1H, ^{19}F, etc., are transferred to dilute spin systems such as ^{13}C or ^{15}N during contact. This process increases the signal-to-noise ratio (S/N) and the signals can be easily identified.

- CP enhances the signal from less abundant nuclei by a factor of γ_I/γ_S, where I is the abundant spin and S is the dilute spin.
- Abundant spins are significantly dipolar coupled and hence have large fluctuating magnetic fields due to motion. As a result, spin-lattice relaxation is fast at the abundant nuclei. Usually, the spin dilute systems

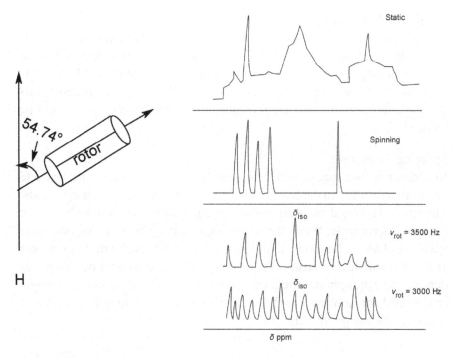

Figure 5.54 Technique, the effect of spinning and spinning rate.

have long relaxation times and hence we have to wait for a long time for them to reach the ground state. This is eliminated by CP because now the relaxation depends the T_1 of the abundant spins (protons, fluorine, etc.).

EXERCISES

Exercise 5.1. NMR is observed in which part of electromagnetic radiation?

Exercise 5.2. While a ^{13}C NMR spectrum can be recorded, ^{12}C NMR spectrum is not observed. Give the reason.

Exercise 5.3. Consider the compound $CH_3CH_2CH_2F$. While recording ^{1}H NMR spectrum, the signal due to F is not seen even though $I \neq 0$ for F. Similarly, while recording ^{19}F NMR spectrum of the compound, signals due to protons are not seen. Explain.

Exercise 5.4. What is the magnetogyric ratio? Explain its importance with examples.

Exercise 5.5. The difference in population of the ground and excited states in ^1H NMR spectroscopy is extremely small. In other words, both the levels are almost equally populated. Hence, after first excitation of the proton from the ground to the excited state, the populations of the two states will become equal and hence further excitation is not possible. Therefore, resonance should not occur further and signals cannot be obtained. However, NMR signals are obtained continuously. How?

Exercise 5.6. What is meant by shielding and deshielding? Explain with suitable examples.

Exercise 5.7. In cyclohexane, there are two types of protons, namely axial and equatorial protons. Which one will appear down field? Why?

Exercise 5.8. The protons attached to N give broad signals. Why?

Exercise 5.9. While the signals of ethylenic protons appear downfield, those of acetylenic protons appear upfield. Why?

Exercise 5.10. Distinguish between chemical and magnetic equivalence with examples.

Exercise 5.11. Diastereotopic and enantiotopic protons behave differently in NMR. Explain.

Exercise 5.12. What is the significance of the coupling constant J in NMR? Explain with examples.

Exercise 5.13. How is NMR useful in identifying the structure of inorganic compounds? Explain with examples.

Exercise 5.14. How is NMR useful in the study of metal complexes? Explain with examples.

Exercise 5.15. State and explain the principle of solid-state NMR?

Exercise 5.16. What is the magic angle in solid-state NMR?

Exercise 5.17. What is cross-polarization? How is it important?

REFERENCES

[1] W. Kemp, Organic Spectroscopy, Palgrave, New York, 2011.

[2] R.M. Silverstein, F.X. Webster, D.J. Kiemle, Spectrometric Identification of Organic Compounds, John Wiley, New Delhi, 2005.

[3] A.K. Brisdon, Inorganic Spectroscopic Methods, Oxford Science, New York, 2010.

[4] R.S. Drago, Physical Methods for Chemists, second ed., Saunders (W.B.), Orlando, 1992.

[5] A.G. Coutsolelos, G.A. Spyroulias, ^{67}Zn NMR, a tool for coordination chemistry problems, in: I.M.Z. Rappoport (Ed.), The Chemistry of Organozinc Compounds, chap. 4, John Wiley, Chichester, 2006.

[6] M.F. Summers, ^{113}Cd NMR spectroscopy of coordination compounds and proteins, Coord. Chem. Rev. 86 (1988) 43–134.

[7] A. Zhdanko, M. Ströbele, M.E. Maier, Coordination chemistry of gold catalysts in solution: a detailed NMR study, Chem. A Eur. J. 18 (46) (2012) 14732–14744.

[8] J.-P. Kintzinger, J.-M. Lahn, E. Kauffmann, J.L. Dyes, A.I. Popov, Anion coordination chemistry. ^{35}Cl NMR studies of chloride anion cryptates, J. Am. Chem. Soc. 105 (1983) 7549–7553.

[9] J.A. Iggo, NMR Spectroscopy in Inorganic Chemistry, Oxford Science Publications, New York, 2011.

INDEX

Note: Page numbers followed by f indicate figures and t indicate tables.

Printed in the United States
By Bookmasters